Principles of Biotechnology

Principles of Biotechnology

Edited by Colin Davenport

SYRAWOOD
PUBLISHING HOUSE

New York

Published by Syrawood Publishing House,
750 Third Avenue, 9th Floor,
New York, NY 10017, USA
www.syrawoodpublishinghouse.com

Principles of Biotechnology
Edited by Colin Davenport

International Standard Book Number: 978-1-68286-594-1 (Hardback)

Cataloging-in-Publication Data

Principles of biotechnology / edited by Colin Davenport.
    p. cm.
Includes bibliographical references and index.
ISBN 978-1-68286-594-1
1. Biotechnology. 2. Genetic engineering. I. Davenport, Colin.
TP248.2 .P75 2018
660.6--dc23

TABLE OF CONTENTS

Preface VII

Chapter 1   **Basics of Biotechnology**.. 1
- Biotechnology 1

Chapter 2   **Cell Expression Systems in Biotechnology** 15
- Cellular Structure 15
- Metabolic Reactions.. 24
- Citric Acid Cycle.. 33
- Anaerobic Oxidation... 43

Chapter 3   **Gene Sequencing: An Integrated Study**............................. 52
- Gene Sequence.. 52
- Competent Cells... 61
- Cloning.. 76
- Polymerase Chain Reaction.. 95
- Cloning Vector 103
- Eukaryotic Vector 108

Chapter 4   **Biotechnology: Methods and Techniques** 112
- Spectroscopy 112
- Electrophoresis.. 158
- Chromatography... 169
- Antibody 181

Chapter 5   **Applied Areas of Biotechnology**.. 195
- Biotechnology in Plant Sciences... 195
- Biotechnology in Animal Breeding.. 197
- Biotechnology in Medicine.. 199

Permissions

Index

PREFACE

Biotechnology is an interdisciplinary field of study which focuses on the development of specified products using living systems or organisms. The recent developments made in the field of biotechnology help our society in developing better health care products and vaccines. Apart from using this technology in the health care sector, it also helps in generating fuel which is less harmful for our environment. In this book, constant effort has been made to make the understanding of the difficult concepts of biotechnology as easy and informative as possible, for the readers. It aims to serve as a resource guide for students and experts alike and contribute to the growth of the discipline.

To facilitate a deeper understanding of the contents of this book a short introduction of every chapter is written below:

Chapter 1- Biotechnology can be defined as a technology that requires organisms to create or amend products for a particular need. Normally, biotechnology is used to enhance the yield gathered from organisms. The importance of water in biological processes and its uses as a reactant is also elucidated here. This chapter on biotechnology offers an insightful focus, keeping in mind the complex subject matter.

Chapter 2- Cells are classified into prokaryotic and eukaryotic cells based on their structure. While prokaryotic cells are unicellular and are deficient of membrane-bound organelles, eukaryotic cells are either single cells or part of a multicellular tissue. In order to completely understand expression system in biotechnology, it is necessary to understand the processes related to it. The following chapter elucidates the varied processes and mechanisms associated with this area of study.

Chapter 3- Gene sequence is arranged randomly and it becomes difficult to separate when they are also unknown. To resolve the problem, two ways have been developed to represent the genomic sequence, i.e. genomic library and cDNA library. Topics such as recombinant DNA and cloning among others have also been discussed. This chapter strategically encompasses and incorporates the major components and key concepts of biotechnology, providing a complete understanding.

Chapter 4- Spectroscopic methods are used to determine protein concentration and estimate DNA and its melting temperature. This section discusses spectroscopic methods and immunological methods as a few procedures under biotechnology. Science and technology have undergone rapid developments in the past decade which has resulted in the discovery of significant tools and techniques in the field of biotechnology; which have been extensively detailed in this chapter.

Chapter 5- Biotechnology has had an effect on human life and its advancement. A few of scientific progresses in the field of biotechnology are genetic engineering, insect control, herbicide resistant plants, resistance protein etcetera. It has further contributed in the field of medicine. Biotechnology's role in medicine which includes production of therapeutically important proteins, gene therapy and monoclonal antibody production among many others are also explored. The aspects elucidated in this chapter are of vital importance, and provide a better understanding of biotechnology.

I owe the completion of this book to the never-ending support of my family, who supported me throughout the project.

Editor

Basics of Biotechnology

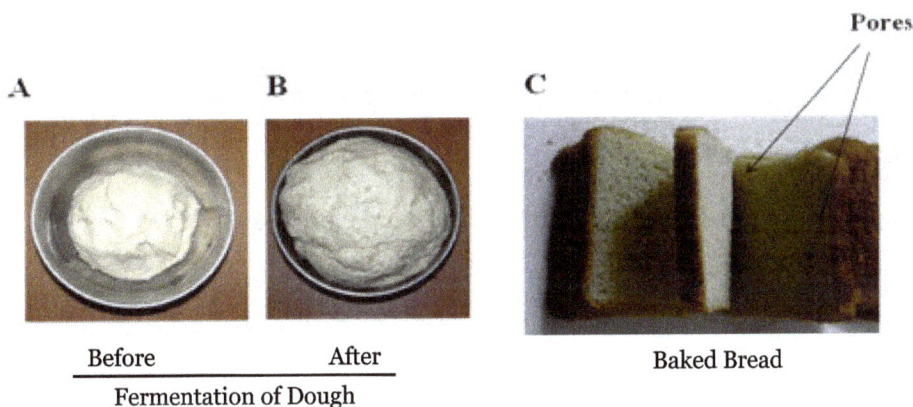

Biotechnology can be defined as a technology that requires organisms to create or amend products for a particular need. Normally, biotechnology is used to enhance the yield gathered from organisms. The importance of water in biological processes and its uses as a reactant is also elucidated here. This chapter on biotechnology offers an insightful focus, keeping in mind the complex subject matter.

Biotechnology

Pores

A   B   C

Before   After   Baked Bread

Fermentation of Dough

Making of Bread from wheat flour. (A) & (B) Dough before and after fermentation.
(C) Cross section of baked Bread. Please note the increase in volume of the dough after fermentation and formation of pores in cross section of bread. Yeast mixed in dough utilizes sugar present in it and produces CO_2 through fermentation, exit of gas causes formation of pores and is responsible for sponginess of bread.

Plant, animal and microbes have been used by humans for nutrition and development of products for consumption such as beer or bread. Understanding of Physical phenomenon has allowed the invention of different types of electronic gadgets, machines, devices and altogether these have been used to increase the efficiency of human activities. Technological advancement has also allowed him to exploit plant, animal and microbial wealth to provide products of commercial or pharmaceutical importance. All these activities (research and development) fall under the big umbrella of biotechnology. In simpler word, Biotechnology is the summation of activities involving technological tools and living organism in such a way that it will enhance the efficiency of the production. The ultimate goal of this field is to improve the product yield from living organism either by employing principles of bio-engineering/bio-process technology or by genetically modifying the organisms. For example, production of bread or other bakery items from wheat flour after adding yeast as fermenting organism. In India, from ancient times wheat flour has been used to prepare "Roti" but yeast has been added to the wheat flour to make it porous by CO_2 generation during fermentation. Since then this process has been very popular in bakery industry and is responsible for prepration of bread, cakes, pizza, etc.

Needs of Biotechnology- The population of india is more than 1 billion and as per projection it may cross 1.5 billion by 2030. This will bring huge burden on biological resources (animal/plant) to provide food for all. Naturally occurring animal, plant or microbial strains have few limitations for them to be utilized for desired products due to following reasons-

1. Purity of the living stock

2. Production of undesired products

3. Secretion of toxic metabolic by-products

4. Inability to withstand harsh biochemical processes/treatments.

5. Higher production cost

6. Susceptible to disease and other environmental conditions

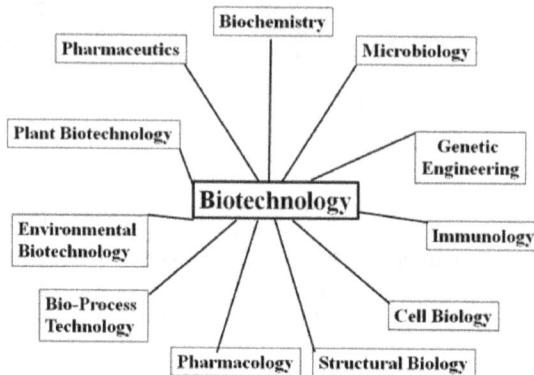

Different Science fields contributing into the advancement of biotechnology

The existing technology today enables us to engineer plants and animals makeing them suitable for maximum production. Living organism has a complex cellular structure, metabolic pathways, genetic make-up, behavior in the synthetic growth media and understanding these processes can help us to modulate specific process/environmental condition or metabolic pathways to achieve the goal of biotechnology. Advancement in different fields of science has paved ways to solve several issues responsible for lower yield of products. Few of the selected science research areas contributing into the development of biotechnology are given in the above figure. The

The foundation of biotechnology relies on the research & development activities in different areas of science and interaction of interdisciplinary areas. The research in the field of plant biotechnology allowed us to produce plants through micro-propagation but with the evident advancement of genetic engineering, it is now possible to produce plant with predefined characteristics imprinted at genetic level through genetic engineering. The similar relationship may also exist for many other overlapping areas and as a result biotechnological operation output is amplified several folds.

Historical Advancement of Biotechnology- Biotechnology related activities depend on two parameters: technological advancement and knowledge of available biota. Technological upgradation goes parallel with the over-all understaning of physical and chemical phenomenon in different time periods. Hence, Biotechnology starts as early as human have realized the importance of organism (animal/plants or microbes) to improve their life-style. A systematic chronological description of biotechnological adavancement over the course of different time periods is given in the table below. The earliest biotechnology related activities are selection and cross breeding of high yielding animals, cross breeding of plants to acquire specific phenotype and preserving the seeds of high yielding crop plant for next sowing season. These were few initial scientific experiments and based on the results, human have made significant modification in available biota. In last century, the systematic and scientific study of living objects with advanced technology has given immense potential to human imagination to either genetically manipulate living organism with desired phenotype or mimic metabolic reactions in an in-vitro system (either in test tube or in cells) to produce molecules with therapeutic importance. Such as "Humulin" is the insulin being produced in bacterial expression system and it is now been making life of millions of diabetic patients easier. Similarly during this era, drought, pest or abiotic resistant plants, high milk yielding animals, transgenic bacteria to produce biofuel, degrade environmental hazard or chelation of heavy metal have been developed. In addition, the historical advancement of biotechnology will not be complete without mentioning development of procedure for artificial insemination and test-tube baby for thousands of couples.

S.No.	Time Period	Major break-through
1.	7000 BC-100CE	• Discovery of fermentation • Crop rotation as a mechanism to improve soil fertility. • Animal and plant products as a source of fertilizer and insecticide respectively.
2.	Pre-20th Century	• Identification of living cell and bacteria • Discovery of small pox vaccine, rabies vaccine. • Process development to separate cream from milk, • Discovery of artificial sweetners, "invertase". • Discovery of DNA and chromosome responsible for genetic traits.

Important milestones of biotechnology

3	20th Century	<ul><li>Discovery of Pencillin.</li><li>3-D Struture of DNA.</li><li>Fabrication of artificial limb and arms,</li><li>Production of human insulin in bacteria "Humulin".</li><li>Discovery of PCR.</li><li>Gene therapy,</li><li>Procedure for artificial insemination and test-tube baby.</li><li>Cloning of first mammal "Dolly".</li></ul>
4	21st Century	<ul><li>Vertebrate, invertebrate and bacterial genome sequences.</li><li>Completion of Human Genome sequence.</li><li>Sequencing of Rice genome.</li><li>Discovery of Nano radio.</li><li>Invention of Bionic leg.</li></ul>

Applications of Biotechnology- Biotechnology has influenced human life in many ways by inventions to make his life more comfortable. Many scientific fields contribute to biotechnology and in return it gives product for their advancement. Few of the biotechnology applications are given in the figure below. The brief description of application of biotechnology in different field is as follows-

Plant sciences- Genetic Engineering has allowed us to produce genetically modified plants with diversified properties such as resistance against pest, drought and abiotic stress. It has enabled us to produce ediable plants with short life-span or ability to grow in different season to increase the number of crops in a year to ultimately increase the food production. Horticulture has used biotechnology tools to produce plants with multiple color, shades, aroma to increase the production of natural colors and scent.

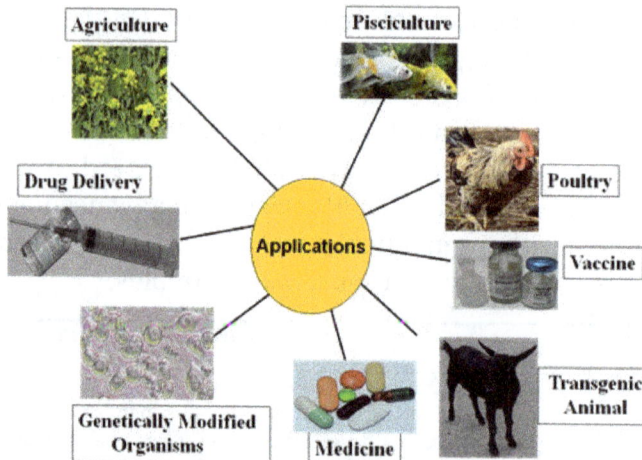

Impact of Biotechnology on different fields & human life.

Animal sciences- One of the early application of biotechnology in animal science is developing method to separate cheese and other food products from milk by enzyme and microbes. Genetic engineering in conjugation with cell biology and biochemistry has developed multiple products of animal origin. Transgenic animal strains with desired phenotype such as high milk yielding animals, fishes and hens with more fat content. A detail description of other biotechnology application in animal sciences is discussed later.

Medicine and Medical Sciences- Biotechnology helped identification of drug like molecules, antibiotics and other medicines. At present a number of antibiotics are being produced by fermentation or in cell based systems. Apart from antibiotic, vaccine, diagnostic kits and other immunotherapy are gift of biotechnological advancement. Development of artificial limb, arms, heart and medical procedures to perform open-heart operation, dialysis, artificial insemination, test-tube baby and other medical procedures.

Biotechnology is the use of living systems and organisms to develop or make products, or "any technological application that uses biological systems, living organisms, or derivatives thereof, to make or modify products or processes for specific use" (UN Convention on Biological Diversity, Art. 2). Depending on the tools and applications, it often overlaps with the (related) fields of bioengineering, biomedical engineering, biomanufacturing, molecular engineering, etc.

For thousands of years, humankind has used biotechnology in agriculture, food production, and medicine. The term is largely believed to have been coined in 1919 by Hungarian engineer Károly Ereky. In the late 20th and early 21st centuries, biotechnology has expanded to include new and diverse sciences such as genomics, recombinant gene techniques, applied immunology, and development of pharmaceutical therapies and diagnostic tests.

Definitions

The wide concept of "biotech" or "biotechnology" encompasses a wide range of procedures for modifying living organisms according to human purposes, going back to domestication of animals, cultivation of the plants, and "improvements" to these through breeding programs that employ artificial selection and hybridization. Modern usage also includes genetic engineering as well as cell and tissue culture technologies. The American Chemical Society defines biotechnology as the application of biological organisms, systems, or processes by various industries to learning about the science of life and the improvement of the value of materials and organisms such as pharmaceuticals, crops, and livestock. As per European Federation of Biotechnology, biotechnology is the integration of natural science and organisms, cells, parts thereof, and molecular analogues for products and services. Biotechnology also writes on the pure biological sciences (animal cell culture, biochemistry, cell biology, embryology, genetics, microbiology, and molecular biology). In many instances, it is also dependent on knowledge and methods from outside the sphere of biology including:

- bioinformatics, a new brand of computer science

- bioprocess engineering

- biorobotics

- chemical engineering

Conversely, modern biological sciences (including even concepts such as molecular ecology) are intimately entwined and heavily dependent on the methods developed through biotechnology and what is commonly thought of as the life sciences industry. Biotechnology is the research and development in the laboratory using bioinformatics for exploration, extraction, exploitation and production from any living organisms and any source of biomass by means of biochemical engineering where high value-added products could be planned (reproduced by biosynthesis, for example), forecasted, formulated, developed, manufactured, and marketed for the purpose of sustainable operations (for the return from bottomless initial investment on R & D) and gaining durable patents rights (for exclusives rights for sales, and prior to this to receive national and international approval from the results on animal experiment and human experiment, especially on the pharmaceutical branch of biotechnology to prevent any undetected side-effects or safety concerns by using the products).

By contrast, bioengineering is generally thought of as a related field that more heavily emphasizes higher systems approaches (not necessarily the altering or using of biological materials *directly*) for interfacing with and utilizing living things. Bioengineering is the application of the principles of engineering and natural sciences to tissues, cells and molecules. This can be considered as the use of knowledge from working with and manipulating biology to achieve a result that can improve functions in plants and animals. Relatedly, biomedical engineering is an overlapping field that often draws upon and applies *biotechnology* (by various definitions), especially in certain sub-fields of biomedical and/or chemical engineering such as tissue engineering, biopharmaceutical engineering, and genetic engineering.

History

Brewing was an early application of biotechnology

Although not normally what first comes to mind, many forms of human-derived agriculture clearly fit the broad definition of "utilizing a biotechnological system to make products". Indeed, the cultivation of plants may be viewed as the earliest biotechnological enterprise.

Agriculture has been theorized to have become the dominant way of producing food since the Neolithic Revolution. Through early biotechnology, the earliest farmers selected and bred the best suited crops, having the highest yields, to produce enough food to support a growing population. As crops and fields became increasingly large and difficult to maintain, it was discovered that specific organisms and their by-products could effectively fertilize, restore nitrogen, and control pests. Throughout the history of agriculture, farmers have inadvertently altered the genetics of their crops through introducing them to new environments and breeding them with other plants — one of the first forms of biotechnology.

These processes also were included in early fermentation of beer. These processes were introduced in early Mesopotamia, Egypt, China and India, and still use the same basic biological methods. In brewing, malted grains (containing enzymes) convert starch from grains into sugar and then adding specific yeasts to produce beer. In this process, carbohydrates in the grains were broken down into alcohols such as ethanol. Later other cultures produced the process of lactic acid fermentation which allowed the fermentation and preservation of other forms of food, such as soy sauce. Fermentation was also used in this time period to produce leavened bread. Although the process of fermentation was not fully understood until Louis Pasteur's work in 1857, it is still the first use of biotechnology to convert a food source into another form.

Before the time of Charles Darwin's work and life, animal and plant scientists had already used selective breeding. Darwin added to that body of work with his scientific observations about the ability of science to change species. These accounts contributed to Darwin's theory of natural selection.

For thousands of years, humans have used selective breeding to improve production of crops and livestock to use them for food. In selective breeding, organisms with desirable characteristics are mated to produce offspring with the same characteristics. For example, this technique was used with corn to produce the largest and sweetest crops.

In the early twentieth century scientists gained a greater understanding of microbiology and explored ways of manufacturing specific products. In 1917, Chaim Weizmann first used a pure microbiological culture in an industrial process, that of manufacturing corn starch using *Clostridium acetobutylicum,* to produce acetone, which the United Kingdom desperately needed to manufacture explosives during World War I.

Biotechnology has also led to the development of antibiotics. In 1928, Alexander Fleming discovered the mold *Penicillium*. His work led to the purification of the antibiotic compound formed by the mold by Howard Florey, Ernst Boris Chain and Norman

Heatley – to form what we today know as penicillin. In 1940, penicillin became available for medicinal use to treat bacterial infections in humans.

The field of modern biotechnology is generally thought of as having been born in 1971 when Paul Berg's (Stanford) experiments in gene splicing had early success. Herbert W. Boyer (Univ. Calif. at San Francisco) and Stanley N. Cohen (Stanford) significantly advanced the new technology in 1972 by transferring genetic material into a bacterium, such that the imported material would be reproduced. The commercial viability of a biotechnology industry was significantly expanded on June 16, 1980, when the United States Supreme Court ruled that a genetically modified microorganism could be patented in the case of *Diamond v. Chakrabarty*. Indian-born Ananda Chakrabarty, working for General Electric, had modified a bacterium (of the *Pseudomonas* genus) capable of breaking down crude oil, which he proposed to use in treating oil spills. (Chakrabarty's work did not involve gene manipulation but rather the transfer of entire organelles between strains of the *Pseudomonas* bacterium.

Revenue in the industry is expected to grow by 12.9% in 2008. Another factor influencing the biotechnology sector's success is improved intellectual property rights legislation—and enforcement—worldwide, as well as strengthened demand for medical and pharmaceutical products to cope with an ageing, and ailing, U.S. population.

Rising demand for biofuels is expected to be good news for the biotechnology sector, with the Department of Energy estimating ethanol usage could reduce U.S. petroleum-derived fuel consumption by up to 30% by 2030. The biotechnology sector has allowed the U.S. farming industry to rapidly increase its supply of corn and soybeans—the main inputs into biofuels—by developing genetically modified seeds which are resistant to pests and drought. By boosting farm productivity, biotechnology plays a crucial role in ensuring that biofuel production targets are met.

Examples

A rose plant that began as cells grown in a tissue culture

Biotechnology has applications in four major industrial areas, including health care (medical), crop production and agriculture, non food (industrial) uses of crops and other products (e.g. biodegradable plastics, vegetable oil, biofuels), and environmental uses.

For example, one application of biotechnology is the directed use of organisms for the manufacture of organic products (examples include beer and milk products). Another example is using naturally present bacteria by the mining industry in bioleaching. Biotechnology is also used to recycle, treat waste, clean up sites contaminated by industrial activities (bioremediation), and also to produce biological weapons.

A series of derived terms have been coined to identify several branches of biotechnology; for example:

- Bioinformatics is an interdisciplinary field which addresses biological problems using computational techniques, and makes the rapid organization as well as analysis of biological data possible. The field may also be referred to as *computational biology*, and can be defined as, "conceptualizing biology in terms of molecules and then applying informatics techniques to understand and organize the information associated with these molecules, on a large scale." Bioinformatics plays a key role in various areas, such as functional genomics, structural genomics, and proteomics, and forms a key component in the biotechnology and pharmaceutical sector.

- Blue biotechnology is a term that has been used to describe the marine and aquatic applications of biotechnology, but its use is relatively rare.

- Green biotechnology is biotechnology applied to agricultural processes. An example would be the selection and domestication of plants via micropropagation. Another example is the designing of transgenic plants to grow under specific environments in the presence (or absence) of chemicals. One hope is that green biotechnology might produce more environmentally friendly solutions than traditional industrial agriculture. An example of this is the engineering of a plant to express a pesticide, thereby ending the need of external application of pesticides. An example of this would be Bt corn. Whether or not green biotechnology products such as this are ultimately more environmentally friendly is a topic of considerable debate.

- Red biotechnology is applied to medical processes. Some examples are the designing of organisms to produce antibiotics, and the engineering of genetic cures through genetic manipulation.

- White biotechnology, also known as industrial biotechnology, is biotechnology applied to industrial processes. An example is the designing of an organism to produce a useful chemical. Another example is the using of enzymes as industrial catalysts to either produce valuable chemicals or destroy hazardous/polluting chemicals. White biotechnology tends to consume less in resources than traditional processes used to produce industrial goods.

The investment and economic output of all of these types of applied biotechnologies is termed as "bioeconomy".

Medicine

In medicine, modern biotechnology finds applications in areas such as pharmaceutical drug discovery and production, pharmacogenomics, and genetic testing (or genetic screening).

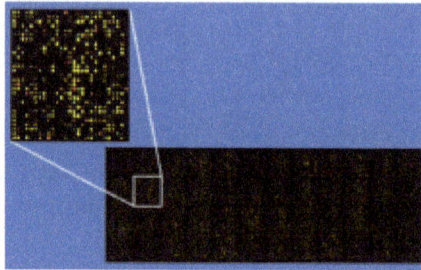

DNA microarray chip – some can do as many as a million blood tests at once

Pharmacogenomics (a combination of pharmacology and genomics) is the technology that analyses how genetic makeup affects an individual's response to drugs. It deals with the influence of genetic variation on drug response in patients by correlating gene expression or single-nucleotide polymorphisms with a drug's efficacy or toxicity. By doing so, pharmacogenomics aims to develop rational means to optimize drug therapy, with respect to the patients' genotype, to ensure maximum efficacy with minimal adverse effects. Such approaches promise the advent of "personalized medicine"; in which drugs and drug combinations are optimized for each individual's unique genetic makeup.

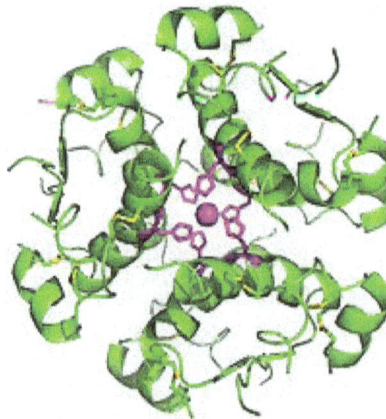

Computer-generated image of insulin hexamers highlighting the threefold symmetry, the zinc ions holding it together, and the histidine residues involved in zinc binding.

Biotechnology has contributed to the discovery and manufacturing of traditional small molecule pharmaceutical drugs as well as drugs that are the product of biotechnology – biopharmaceutics. Modern biotechnology can be used to manufacture existing medicines relatively easily and cheaply. The first genetically engineered products were medicines designed to treat human diseases. To cite one example, in 1978 Genentech developed synthetic humanized insulin by joining its gene with a plasmid vector inserted into the bacterium *Escherichia coli*. Insulin, widely used for the treatment of diabetes, was previously extracted from the pancreas of abattoir animals (cattle and/or pigs). The resulting genetically engineered bacterium enabled the production of vast quantities of synthetic human insulin at relatively low cost. Biotechnology has also enabled emerging therapeutics like gene therapy. The application of biotechnology to basic science (for example through the Human Genome Project) has also dramatically improved our understanding of biology and as our scientific knowledge of normal and disease biology has increased, our ability to develop new medicines to treat previously untreatable diseases has increased as well.

Genetic testing allows the genetic diagnosis of vulnerabilities to inherited diseases, and can also be used to determine a child's parentage (genetic mother and father) or in general a person's ancestry. In addition to studying chromosomes to the level of individual genes, genetic testing in a broader sense includes biochemical tests for the possible presence of genetic diseases, or mutant forms of genes associated with increased risk of developing genetic disorders. Genetic testing identifies changes in chromosomes, genes, or proteins. Most of the time, testing is used to find changes that are associated with inherited disorders. The results of a genetic test can confirm or rule out a suspected genetic condition or help determine a person's chance of developing or passing on a genetic disorder. As of 2011 several hundred genetic tests were in use. Since genetic testing may open up ethical or psychological problems, genetic testing is often accompanied by genetic counseling.

Agriculture

Genetically modified crops ("GM crops", or "biotech crops") are plants used in agriculture, the DNA of which has been modified with genetic engineering techniques. In most cases the aim is to introduce a new trait to the plant which does not occur naturally in the species.

Examples in food crops include resistance to certain pests, diseases, stressful environmental conditions, resistance to chemical treatments (e.g. resistance to a herbicide), reduction of spoilage, or improving the nutrient profile of the crop. Examples in non-food crops include production of pharmaceutical agents, biofuels, and other industrially useful goods, as well as for bioremediation.

Farmers have widely adopted GM technology. Between 1996 and 2011, the total surface area of land cultivated with GM crops had increased by a factor of 94, from 17,000 square kilometers (4,200,000 acres) to 1,600,000 km² (395 million acres). 10% of the world's crop lands were planted with GM crops in 2010. As of 2011, 11 different transgenic crops were grown commercially on 395 million acres (160 million hectares) in 29 countries such as the USA, Brazil, Argentina, India, Canada, China, Paraguay, Pakistan, South Africa, Uruguay, Bolivia, Australia, Philippines, Myanmar, Burkina Faso, Mexico and Spain.

Genetically modified foods are foods produced from organisms that have had specific changes introduced into their DNA with the methods of genetic engineering. These techniques have allowed for the introduction of new crop traits as well as a far greater control over a food's genetic structure than previously afforded by methods such as selective breeding and mutation breeding. Commercial sale of genetically modified foods began in 1994, when Calgene first marketed its Flavr Savr delayed ripening tomato. To date most genetic modification of foods have primarily focused on cash crops in high demand by farmers such as soybean, corn, canola, and cotton seed oil. These have been engineered for resistance to pathogens and herbicides and better nutrient profiles. GM livestock have also been experimentally developed, although as of November 2013 none are currently on the market.

There is a scientific consensus that currently available food derived from GM crops poses no greater risk to human health than conventional food, but that each GM food needs to be tested on a case-by-case basis before introduction. Nonetheless, members of the public are much less likely than scientists to perceive GM foods as safe. The legal and regulatory status of GM foods varies by country, with some nations banning or restricting them, and others permitting them with widely differing degrees of regulation.

GM crops also provide a number of ecological benefits, if not used in excess. However, opponents have objected to GM crops per se on several grounds, including environmental concerns, whether food produced from GM crops is safe, whether GM crops are needed to address the world's food needs, and economic concerns raised by the fact these organisms are subject to intellectual property law.

Industrial

Industrial biotechnology (known mainly in Europe as white biotechnology) is the application of biotechnology for industrial purposes, including industrial fermentation. It includes the practice of using cells such as micro-organisms, or components of cells like enzymes, to generate industrially useful products in sectors such as chemicals, food and feed, detergents, paper and pulp, textiles and biofuels. In doing so, biotechnology uses renewable raw materials and may contribute to lowering greenhouse gas emissions and moving away from a petrochemical-based economy.

Environmental

The environment can be affected by biotechnologies, both positively and adversely. Vallero and others have argued that the difference between beneficial biotechnology (e.g. bioremediation to clean up an oil spill or hazard chemical leak) versus the adverse effects stemming from biotechnological enterprises (e.g. flow of genetic material from transgenic organisms into wild strains) can be seen as applications and implications, respectively. Cleaning up environmental wastes is an example of an application of environmental biotechnology; whereas loss of biodiversity or loss of containment of a harmful microbe are examples of environmental implications of biotechnology.

Regulation

The regulation of genetic engineering concerns approaches taken by governments to assess and manage the risks associated with the use of genetic engineering technology, and the development and release of genetically modified organisms (GMO), including genetically modified crops and genetically modified fish. There are differences in the regulation of GMOs between countries, with some of the most marked differences occurring between the USA and Europe. Regulation varies in a given country depending on the intended use of the products of the genetic engineering. For example, a crop not intended for food use is generally not reviewed by authorities responsible for food safety. The European Union differentiates between approval for cultivation within the EU and approval for import and processing. While only a few GMOs have been approved for cultivation in the EU a number of GMOs have been approved for import and processing. The cultivation of GMOs has triggered a debate about coexistence of GM and non GM crops. Depending on the coexistence regulations incentives for cultivation of GM crops differ.

Learning

In 1988, after prompting from the United States Congress, the National Institute of General Medical Sciences (National Institutes of Health) (NIGMS) instituted a funding mechanism for biotechnology training. Universities nationwide compete for these funds to establish Biotechnology Training Programs (BTPs). Each successful application is generally funded for five years then must be competitively renewed. Graduate students in turn compete for acceptance into a BTP; if accepted, then stipend, tuition and health insurance support is provided for two or three years during the course of their Ph.D. thesis work. Nineteen institutions offer NIGMS supported BTPs. Biotechnology training is also offered at the undergraduate level and in community colleges.

References

- Braun, Charles L.; Smirnov, Sergei N. (1993-08-01). "Why is water blue?". Journal of Chemical Education. 70 (8): 612. Bibcode:1993JChEd..70..612B. doi:10.1021/ed070p612. ISSN 0021-9584

- Riddick, John (1970). Organic Solvents Physical Properties and Methods of Purification. Techniques of Chemistry. Wiley-Interscience. Table of Physical Properties, Water ob. pg 67-8. ISBN 0471927260

- "Can the ocean freeze?". National Ocean Service. National Oceanic and Atmospheric Administration. Retrieved 2016-06-09

- Nomenclature of Inorganic Chemistry: IUPAC Recommendations 2005 (PDF). Royal Society of Chemistry. 22 Nov 2005. p. 85. ISBN 978-0-85404-438-2. Retrieved 2016-07-31

- Fine, R.A. & Millero, F.J. (1973). "Compressibility of water as a function of temperature and pressure". Journal of Chemical Physics. 59 (10): 5529. Bibcode:1973JChPh..59.5529F. doi:10.1063/1.1679903

- "Experimenter Drinks 'Heavy Water' at $5,000 a Quart". Popular Science Monthly. 126 (4). New York: Popular Science Publishing. Apr 1935. p. 17. Retrieved 7 Jan 2011

- Zumdahl, Steven S.; Zumdahl, Susan A. (2013-01-01). Chemistry (9th ed.). Cengage Learning. p. 981. ISBN 978-1-13-361109-7

- "Revised Release on the Pressure along the Melting and Sublimation Curves of Ordinary Water Substance" (PDF). IAPWS. September 2011. Retrieved 2013-02-19

- Lewis, G. N.; MacDonald, R. T. (1933). "Concentration of H2 Isotope". The Journal of Chemical Physics. 1 (6): 341. Bibcode:1933JChPh...1..341L. doi:10.1063/1.1749300

- Leigh, G. J.; et al. (1998). Principles of chemical nomenclature: a guide to IUPAC recommendations (PDF). Blackwell Science Ltd, UK. p. 34. ISBN 0-86542-685-6. Archived from the original (PDF) on 2011-07-26

- Crofts, A. (1996). "Lecture 12: Proton Conduction, Stoichiometry". University of Illinois at Urbana-Champaign. Retrieved 2009-12-06

- Review of the vapour pressures of ice and supercooled water for atmospheric applications. D. M. Murphy and T. Koop (2005) Quarterly Journal of the Royal Meteorological Society, 131, 1539

- Campbell, Neil A.; Brad Williamson; Robin J. Heyden (2006). Biology: Exploring Life. Boston, Massachusetts: Pearson Prentice Hall. ISBN 0-13-250882-6

- Müller, Grover C. (June 1937). "Is 'Heavy Water' the Fountain of Youth?". Popular Science Monthly. 130 (6). New York: Popular Science Publishing. pp. 22–23. Retrieved 7 Jan 2011

- Sharp, Robert Phillip (1988-11-25). Living Ice: Understanding Glaciers and Glaciation. Cambridge University Press. p. 27. ISBN 0-521-33009-2

- "Joseph Louis Gay-Lussac, French chemist (1778–1850)". 1902 Encyclopedia. Footnote 122-1. Retrieved 2016-05-26

2

Cell Expression Systems in Biotechnology

Cells are classified into prokaryotic and eukaryotic cells based on their structure. While prokaryotic cells are unicellular and are deficient of membrane-bound organelles, eukaryotic cells are either single cells or part of a multicellular tissue. In order to completely understand expression system in biotechnology, it is necessary to understand the processes related to it. The following chapter elucidates the varied processes and mechanisms associated with this area of study.

Cellular Structure

Feature	Prokaryote	Eukaryote
Size	Small, in μm range	Variable size, upto 40μm in diameter.
Genetic material	Circular DNA present in cytosol as free material	DNA in the form of linear chromosome present in well defined double membrane nucleus, no direct connection with cytosol
Replication	Single origin of replication	Multiple origin of replication.
Genes	No Intron	Presence of Intron
Organelles	No membrane bound organelles	Membrane bound orgelles with well defined function.
Cell walls	Very complex cell wall	Except Fungi and plant, eukaryotic cells are devoid of a thick cell wall.
Ribosome	70S	80S
Trancription and translation	Occurs together	Transcription in nucleus and translation in cytosol

Different between Prokaryotic and Eukaryotic cells

Higher eukaryotes have multiple organs to perform specific functions such as liver, kidney and heart. Each Organ has specific tissue and each tissue is composed of cells. "Cell is the structural and functional unit of life" and it contains all necessary infrastructure to peform all functions. Based on cellular structure, cells are classified as prokaryotic and eukaryotic cells. In most of the cases, prokaryotes are single cells where as eukaryotes are either single cells or part of multicellular tissues system. Besides this, both types of cells have several structural and metabolic differences.

Structure of Prokaryotic cells- A prokaryotic cell is much simpler and smaller than eukarotic cells. It lacks membrane bound organelles including nucleus. A typical prokaryotic cells is shown in the figure below. The description of different structural feature of prokaryotic cells is as follows-

1. Outer Flagella: A flagellum attached to the bacterial capsule is a central feature of most of the prokaryotic cell especially of the motile bacteria. It provides motion or locomotion to the bacteria and be responsible for chemotaxis of bacteria. Movement of bacteria towards a chemical gradient (such as glucose) is known as chemotaxis. Flagellum is a part of cell wall and its motion is regulated by motor proteins present inside the cell. Flagellar motion is an energy consuming process and it is governed by an ATPase present at the bottom of the shaft. It is made up of protein flagellin and reduction or suppression of flagellar protein reduces bacterial infectivity (pathogenicity) and ability to grow.

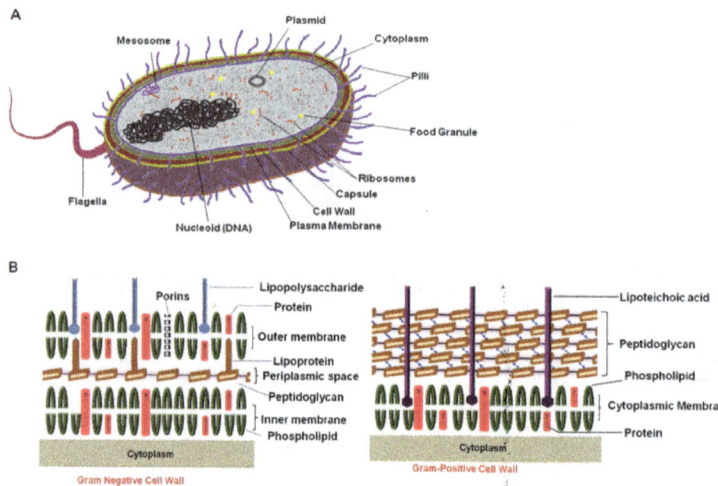

Structural details of a typical prokaryotic cell. (A) Whole cell and (B) composition of cell wall of gram negative and positive bacteria.

2. Bacterial surface layers: Bacteria posses 3 anatomical barriers to protect the cells from external damage. Bacterial capsule is the outer most layer and made up of high molecular weight polysaccharides. It is impermeable to the water or other aqueous solvent and it is responsible for antigenicity of bacterial cells. Cell wall in bacteria and its response to gram staining is the basis of classification of bacterial species.

Cell wall composition in gram-ve and gram +ve bacteria is different. Bacterial cell wall has different constituents and be responsible for their reactivity towards gram stain.

A. Peptidoglycan layer: peptidoglycan layer is thick in gram +ve bacteria and thin in gram –ve bacteria. Peptidoglycan is a polymer of NAG (N-acetyl-glucosamine) and NAM (N-acetyl-muramic acid) linked by a β-(1,4) linkage. Sugar polymer are attached to peptide chain composed of amino acids, L-alanine, D-glutamic acid, L-lysine and D-alanine. Peptide chain present in one layer cross linked to the next layer to form a mesh work and be responsible for physical strength of the cell wall. Peptidoglycan synthesis is targeted by antibiotics such as pencillin where as lysozyme (present in human saliva or tears) degrades the peptidoglycan layer by cleaving glycosidic bond connecting NAG-NAM to form polymer.

B. Lipoteichoic acids : Lipoteichoic acid (LTA) are only found in gram +ve bacteria cell wall and it is an important antigenic determinant.

C. Lipopolysaccharides (LPS)- Lipopolysaccharides (LPS) are found only in gram –ve bacterial cell wall and it is an important antigenic determinant.

3. Cytosol and other organelles- Prokaryotic cells do not contains any membrane bound organelle. The organelles are present in cytosol such as ribosome (70S), genetic material where as electron transport chain complexes are embedded within the plasma membrane.

4. Chromosome and extra chromosomal DNA- Prokaryote cell contains genetic material in the form of circular DNA, known as " bacterial chromosome ". It contains genetic elements for replication, transcription and translation. Bacterial chromosome follows a rolling circle mode of DNA replication. The genes present on chromosome does not contains non coding region (introns) and it is co-translated to protein. Besides main circle DNA, bacteria also contains extra chromosomal circular DNA known as "plasmid". Presence of plasmid containing resistance gene confers resistance towards known antibiotics. Exchange of extra-chromosomal DNA between different bacterial strains is one of the mechanisms responsible for spread of antibiotic resistance across the bacterial population.

Struture of Eukaryotic cell- The eukaryotic cell is much more complex and it contains many membrane bound organelles to perform specific functions. It contains a nucleus isolated from cytosol and enclosed in a well defined double membrane. A typical eukaryotic animal and plant cell is shown in the figure below.

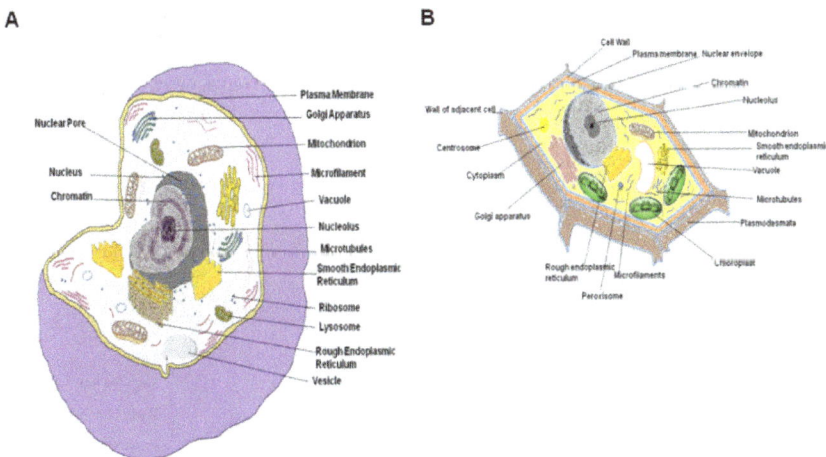

Structure of Eukaryotic cell. (A) Animal Cell (B) Plant Cell

FEATURE	PLANT CELL	ANIMAL CELL
Cell wall	Present	Mostly absent
Size	Large	Comparatively small
Chlorophyll	Present	Absent
Vacuole	Large Central	Small and many in number
Mitochondria	Few	More
Lysosome	Almost absent	Present
Glyoxysomes	Present	Absent
Cytokinesis	By Plate method	By constriction

Difference between Animal and Plant cells

The description of different structural feature of eukaryotic cell is as follows-

Different organelles of Eukryotic cells (Animal)

1. Cytosol- Cytosol is the liquid part filled inside the cell and it contains water, salt, macromolecules (protein, lipid, RNA). It has an array of microtubule fiber running through out the cytosol to give vesicular structure to its destination. Besides this, cytosol exhibits "Sol" to "Gel" transition and such transition regulates multiple bio-chemical and cellular processes.

2. Nucleus- Nucleus is the central processing unit of cell and homologous to the processor in a typical computer. The liquid filled inside nucleus is called as nucleoplasm. It is a viscous liquid containing nucleotides and enzymes to perform replication, transcription, DNA damage repair etc. It contains genetic material (DNA) in a complex fashion involving several proteins (histones) to pack into nuclear bodies or chromosomes. The chromatin in eukarotic nucleus is divided into euchromatin or heterochromatin. Euchromatin is a part of chromatin where DNA is loosely packed and it is transcriptionally active to form mRNA where as Heterochromatin is more densily packed and it is transcriptionally inactive. Nuclei in eukarytotic cells are present in a double layer of membrane known as nuclear envelope. Outer membrane of nuclear envelope is continuous with the rough endoplasmic reticulum and has ribosome attached to it. The space between these two membranes is called as perinuclear space. Nuclear envelope often has nuclear pore and as per calculation an average nucleus has 3000-4000 pores per nuclear envelope.

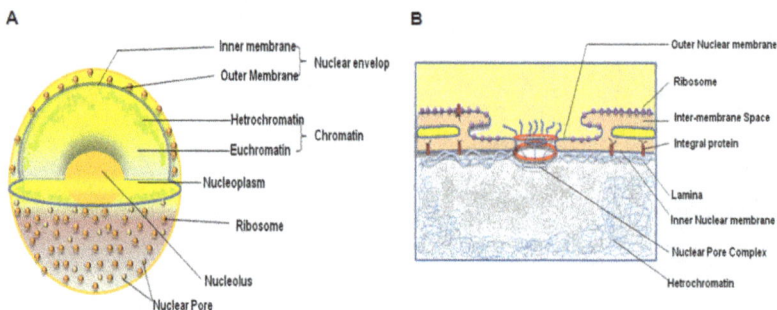

Structural details of nucleus. (A) whole and (B) enlarged view of nuclear pore.

Nuclear pore is 100nm is diameter and consists of several proteins. It is a gateway for transfer of material between nucleus and cytosol. RNA formed after transcription from DNA within the nucleus and move out of the nucleus into the cytosol through nuclear pore. Similarly protein from cytosol crosses nuclear pore to initiate replication, transcription and other processes.

1. Mitochondria- It is popularly known as " power house of the cell " as the organelle is actively involved in the generation of ATP to run the cellular activities. Mitochondria is a double layered membrane-bound organelle with different structural properties. Outer membrane is smooth and cover the complete organelle with large number of integral proteins, known as porins.

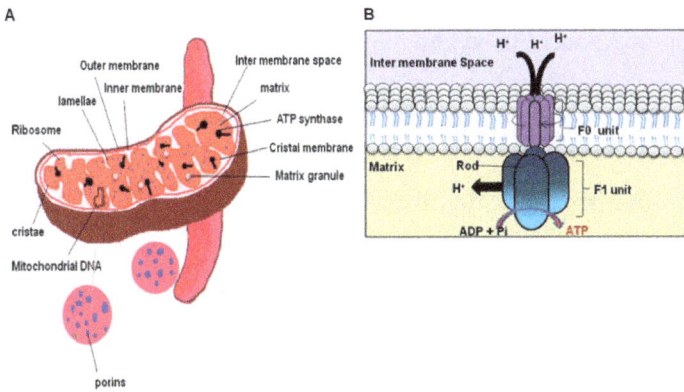

Mitochondria. (A) Struture of mitochondria and (B) enlarged view of ATP Synthase.

Porin allows free movement of molecules less than 5000da within and outside mitochondria. Where as larger molecules or proteins moves into the mitochondria through transporters involving signal peptides known as " mitochondrial targeting sequence ". Inner membrane is folded into membrane projections to form cristae.

Cristae occupies major area of membrane surface and house machinery for anaerobic oxdidation and electron transport chain to produce ATP. Due to presence of inner and outer membrane, mitochondria can be divided into 2 compartments: first in between the inner and outer membrane, known as intermembrane space and second inside the inner membrane known as matrix . The proteins present in intermembrane space have a role in executing " programmed cell death " or " apoptosis ". Matrix is the liquid part present in the inner most compartment of the mitochondria and it contains ribosome, DNA, RNA, enzymes to run Kreb's cycle and other proteins. Mitochondrial DNA is circular and it has full machinery to synthesize its own RNA (mRNA, rRNA and t-RNA) and proteins. Marked differences exist between mitochondrial DNA and DNA present in nucleus and these differences are not discussed here due to space constrain. Electron transport chain components (complex I to complex V) are integral proteins, present in the inner membrane of mitochondria. During metabolic reactions such as glycolysis, Kreb's cycle [metabolic reaction are discussed later] produces large amount of reducing equivalent in the form of $NADH_2$ and $FADH_2$. Electron transport chain process reducing equivalent and flow of the electron through different complexes (Complex I to Complex IV) causes generation of proton gradient across the membrane. Proton expelled in the intermembrane space returned back to the matrix through complex V (ATP synthase) to generates ATP. ATP synthase is a mushroom shaped multimeric protein complex, mainly composed of two proteins F_o and F_1. F_o is a membrane bound portion where as F_1 is the complex present into the lumen towards matrix. F_oF_1 complex of mitochondria harvest the proton motive force to catalyze phosphorylation reaction involving ADP and phosphate to generate ATP.

Functions of mitochondria-

1. Production of ATP

2. Generation of Reactive Oxygen Species (ROS) in immune cells to kill infectious agents.

3. Used to track tree of a family.

4. Role in programmed cell death or "apoptosis"

2. Chloroplast- Chloroplasts are found in plant, algae and other lower invertebrates such as euglena. Contrasting to mitochondria, chloroplast has outer membrane, an inner membrane and then light pigment containing inner most thylakoid membrane. Outer membrane is porous to the small molecules but protein or large molecules are transported by TOC (translocon on the outer chloroplast membrane) complex. Movement of material passed through outer membrane gets into the inner membrane through TIC (translocon on the inner chloroplast membrane) complex. In between outer and inner membrane is intermembrane space filled with aqueous liquid.

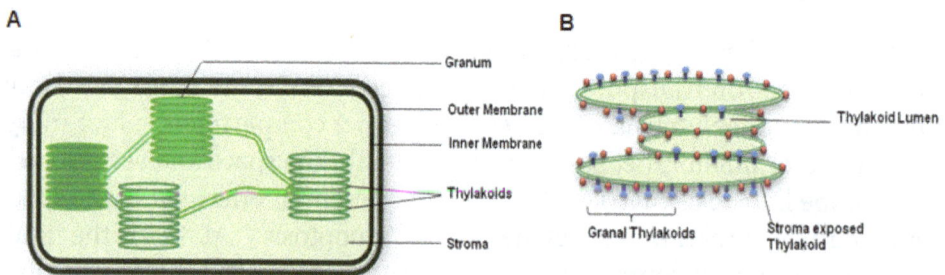

(A) Struture of Chloroplast, (B) Arrangement of thylakoid membrane in chloroplast.

The inner membrane of the chloroplast further folds to a flattend membrane system known as thylakoids . The photosynthsis machinery such as light absorbing pigments, electron carriers and ATP synthesizing machinery is present on inner membrane as intergral protein complex. Thylakoid membranes are arranged like stack of coin to form granum. The granum throughout the chloroplast are connected by tubule to share the material. Over-all structure of chloroplast is similar to mitochondria but it has few significant structural and biochemical differences. Thylakoid membrane contains photo-synthetic green colored pigment chlorophyll.

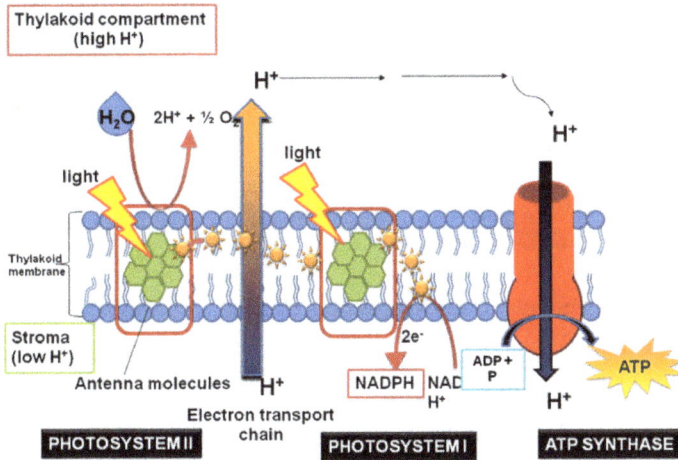

$$6CO_2 + 6H_2O + Solar\ Energy \rightarrow C_6H_{12}O_6 + 6O_2$$

Different Steps of Photosynthesis.

Photosynthesis is an assimilation reaction involving CO_2 and water to produce sugar in the presence of solar energy (photons) that catalyzes fusion reaction as given Eq. 4.1. The photo system present on thylakoid membrane consists of two photo system, photo system-I (PS-I) and photo system complex II (PS-II) . PS-II absorbs the photon from solar energy to excite the electron to the higher energy state, and catalyze water break down into the proton and oxygen. The electron pass through multiple electron carrier and during this proton are exported out of the thylakoid membrane into the lumen. The proton passes through ATP synthase and returns back into the stroma to generate ATP. The electron from PS-II is eventually been received by PS-I and been excited after absorbing photon from sun light to high energy state. The energy associated with these electrons are used to generate NADPH in the stroma. Hence as a result of photosynthesis, solar energy is trapped by photo synthesis apparatus to generate ATP and NADPH into the lumen. Both of them are used to run Calvin cycle to assimilate environmental CO_2 to form sugar.

3. Organelles of Vesicular Trafficking System: The main function of these organelles is to manage the distribution of material (food particles or proteins) throughout the cell. 3 different organelles such as endoplasmic reticulum, Golgi apparatus and lysosome, co-ordinately work together to maintains vesicular transport of material across the cell. Eukaryotic cell takes up the solid material from outside through a process called " endocytosis " whereas uptake of liquid is through a process called as " pinocytosis ". Similarly material is secreted out of the cells through "exocytosis". In addition, intravesicular system delivers protein synthesized in endoplasmic reticulum to different organelles.

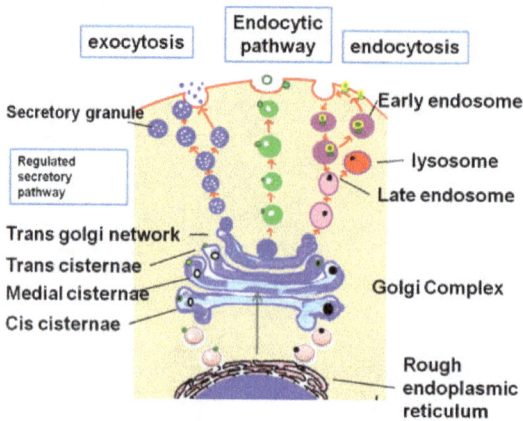

Intra cellular vesicular trafficking system of cell.

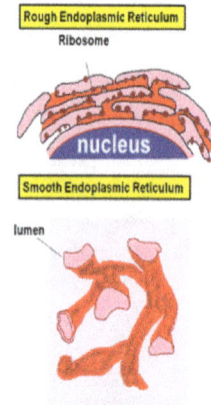

Endoplasmic reticu-
lum.

During endocytosis, material present outside the cells binds to the cells surface through cell surface receptors and trap it in a membraneous structure called as endosome . Endosomal vesicles are fused with the lysosomes to form late endosome. In late endosome, with the help of lysosomal enzymes material is digested and then endosome is fused with the Golgi bodies and deliver the content for further distribution. In the similar manner, during secretion, vesicles originate from Golgi bodies and fuse with the plasma membrane to release the content outside of the cell.

4. Endoplasmic Reticulum- The vesicular network starts from nuclear membrane and spread throughout the cytosol constitutes endoplasmic reticulum. There are two different types of endoplasmic reticuli present in the cell, 1) Rough endoplasmic reticulum (RER), and 2) smooth endoplasmic reticulum (SER). RER has ribosome attached to it to give a rough appearance whereas smooth endoplasmic reticulum is devoid of ribosomes. Protein synthesis on ribosome attached to RER are sorted into 3 different catagories, such as integral membrane proteins, proteins for secretion and protein destined for different organelles. Proteins are synthesized with a n-signal peptide and these signal peptides are recognized by signal recognition particle on their the target organelles. For example, if a protein is synthesized with a signal peptide for mitochondria, it will attach to signal recognition particle and receptor onto the outer mitochondrial membrane to deliver the protein. The proteins without any signal peptide tags are supposed to remain in the cytsol.

Functions Of Endoplasmic Reticulum:

1. Synthesis of steroid hormone in gonad cells.

2. Detoxification

3. Ca^{2+} sequestration

4. Synthesis of protein, phospholipid and carbohydrate.

5. Protein sorting to different organelles.

6. Protein modifications such as glycosylation etc.

Golgi Bodies- Golgi bodies were first visualized by a metallic stain invented by Camillo golgi and it is made of flattend, disk like cisternae arranged in a stacked manner to give 3 distinct zones. Cis-face receives material or vesicles from endoplasmic reticulum, medial Golgi is the actual place where protein are covalently modified with the sugar. Trans Golgi is the face of Golgi towards plasma membrane and this site sorts vesicle for their destined organelles or plasma membrane.

Functions of Golgi Bodies

1. Protein sorting

2. Protein modifications (Glycosylation)

3. Proteolysis

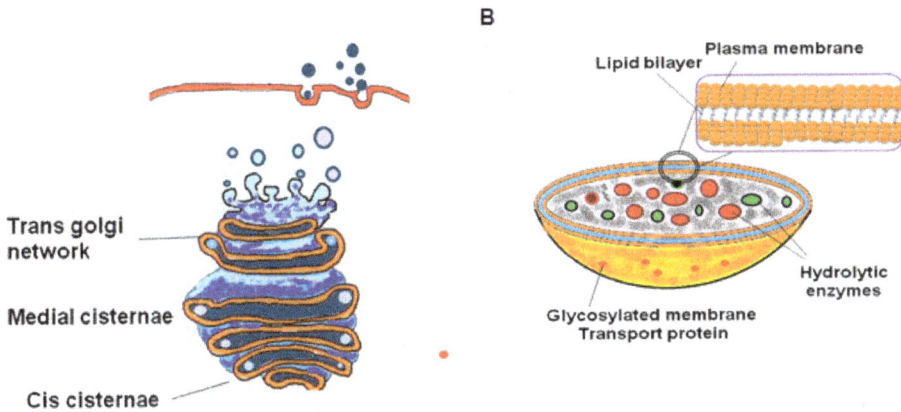

Scematic structure of (A) Golgi bodies and (B) Lysosome.

Lysosomes- Lysosomes are discovered by De Duve. They are membrane bound organelles and an important component of intracellular vesicular system. They are popularly known as suicidal bags due to their role in autophagy, a cellular process probably operates in cells during starvation to meet their energy requirements. Lysosome lumen is extremely acidic and contains protease, cytolytic enzymes to degrade the ingested material.

Functions of Lysosomes

1. Degradation of ingested food material for delivery through vesicular system.

2. Degradation of pathogenic bacteria

3. Degradation of old protein.

Metabolic Reactions

Cellular integrity is maintained at the expense of energy produced by a set of chemical reactions, collectively known as metabolism. It is a summation of two different types of chemical processes:

Anabolism, the reactions which are responsible for formation of new compounds. It is alternatively known as biosynthetic pathway.

Catabolism, the reactions which are responsible for utilization of organic nutrients to produce energy in the form of ATP, NADH, FADH. ATP is the readily available form of energy whereas NADH and FADH needs to go mitochondria for ATP generation. Although carbohydrate, protein and fat undergo catabolism to produce energy but carbohydrate is most preferred choice for this purpose and henceforth topic of choice to discuss in the current course.

Carbohydrate Metabolism- Post digestion, food material is digested into the amino acid, fatty acid and glucose. All these final digestion products are absorbed by intestine and enter into the blood stream. Glucose enters into blood and distribute to the different organs for storage purpose but liver is the prime site for storage. Glucose is converted into the glycogen with the help of an enzyme glycogen synthase. Glucose is oxidized into the glycolysis and Kreb's cycle to produce ATP and other reducing equivalent to produce energy.

Glycolysis- Glycolysis is central to carbohydrate metabolism and it is the universal pathway found in prokaryotic or eukaryotic cells. It is a breakdown of 6 membered glucose into two 3 membered carbon suger to feed Kreb's cycle (in the presence of oxygen) or to send for anaerobic oxidation (in the absence of oxygen). Hence, it plays a crucial role for adopation of a living organism under differet types of stress conditions. The glycolysis is a 10 step chemical reaction to enable glucose for its optimal oxidation. All these reactions are given in the figure.

STEP-1: Phosphorylation of glucose- Glucose produced after glycogen breakdown is phosphorylated by glucokinase (in liver) or hexokinase in all other tissues especially in muscles. In the phosphorylation reaction, phosphate (γ-phosphate) group of ATP is transferred to glucose to form glucose-6-phosphate. The phosphorylation reaction of glucose to produce glucose-6-phosphate marks the molecule for glycolysis. One molecule of ATP is utilized in this step.

STEP 2: Conversion of glucose-6-phosphate to fructose-6-phosphate- Phosphorylated sugar produced in step-1 is converted into the fructose-6-phosphate by the action of phospho-hexose isomerase.

STEP 3: Phosphorylation of fructose-6-phosphate- In this step, sugar is further phsophorylated at carbon 1 to produce fructose-1, 6 bis phosphate by the action of

Phosphofructokinase. In the phosphorylation reaction, phosphate (γ-phosphate) group of ATP is transferred to phosphorylated sugar to form fructose-1, 6 bis phosphate. One molecule of ATP is utilized in this step.

STEP 4: Clevage of fructose 1, 6-bis phosphate- This step is catalyzed by enzyme aldolase or fructose 1, 6 bis aldolase to generate glyceraldehyde-3 phosphate (aldose) and dihydroxy acetone phosphate (ketose).

STEP 1-4: First 4 reactions of enzymatic conversion of glucose (6 carbon sugar) to glyceraldehydes-3 phosphate (aldose) and dihydroxy acetone phosphate (ketose) are considered as preparative phase of glycolysis and during this phase, two major events happen:

1. Commitment of Sugar for glycolysis- Phosphorylated products are negatively charged and impermeable to the cell membrane through passive diffusion. Glycolysis operates in cytosol and as a result first step of phosphorylation inhibits the passive movement of the particular glucose moiety and drive it to participate in further steps of glycolysis.

2. Activation of sugar- In the 1st and 3rd step of glycolysis, two phosphorylation reactions add potential energy into the molecule and hence activate the sugar to participate into the cleavage reaction to form two 3 carbon sugar moiety.

STEP 5: Interconversion of the triose phosphates- Three carbon sugar formed in step 4 undergoes internal conversion and as glyceraldehyde-3 phosphate can readily be able to enter into the next step, the ketose generated in step 4 is reversibly convereted into the glyceraldehydes-3 phosphate by triose-3-phosphate isomerase.

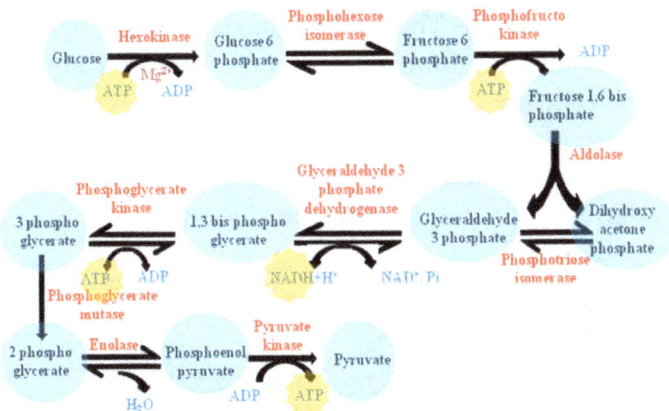

Different Reactions of Glycolysis.

STEP 6: Glyceraldehyde-3-phosphate to 1,3 bis-phospho-glycerate- In this step, one molecule of NADH is produced after oxidation of aldehyde group of glyceraldehyde-3-phosphate with the help of enzyme glyceraldehyde-3-phosphate dehydrogenase.

STEP 7: In this step, phosphate group from 1,3 bis-phosphoglycerate is removed by phosphoglycerate kinase with an acyl phosphate group transfer to ADP to generate ATP molecule.

STEP 8: Conversion of 3-phosphoglycerate to 2-phosphoglycerate- In a two step mechanism, phosphoglycerate mutase catalyzes a reversible shift of phosphoryl group to form 2-phosphoglycerate.

STEP 9: Dehydration of 2-phosphoglycerate to phosphoenol pyruvate- The enzyme enolase catalyzes the dehydration reaction to produce phosphoenol pyruvate , a compound with high phosphoryl group transfer potential.

STEP 10: In the last step of glycolysis, phosphate group from phosphoenol pyruvate is transferred by pyruvate kinase with an acyl phosphate group transfer to ADP to generate ATP molecule.

Calculation of ATP production during Glycolysis

The balance sheet of ATP generation from one molecule of glucose is as follow-	
STEPS OF GLYCOLYSIS	Number of ATP Generation (+) or Investment (-)
1. Step 1-4	- 2
2. Generation of 2 molecules of glyceraldehyde-3 phosphate.	
3. Step 6, generation of NADH, Each NADH in ETS gives 3 ATP	2x3=6
4. Step 7, Generation of ATP	2x1=2
5. Step 10, Generation of ATP	2x1=2
NET BALANCE for oxidation of one glucose molecule.	6+2+2-2= 8 ATP molecules

Regulation of Glycolysis-

1. Uptake of glucose from blood- The level of glucose present in a cell determines the availability of sugar for oxidation via glycolysis. Glucose transport in cell is regulated by several cell surface receptor which are under the control of insulin. Insulin upregulates the level of glucose transporters Glut-3 or Glut-4 and increases the uptake of glucose from blood stream. In addition, insulin also regulates breakdown of glycogen to increase the amount of available glucose.

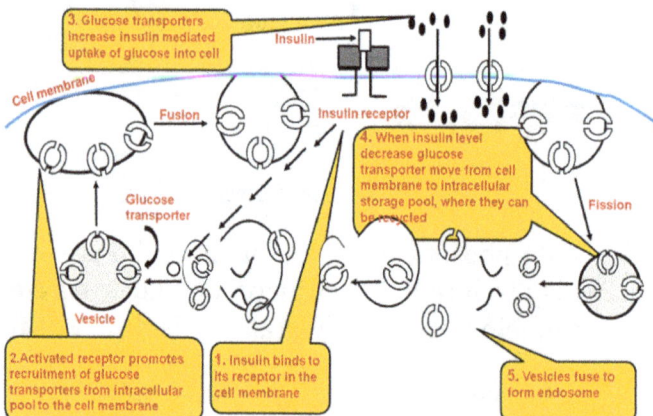

Regulation of uptake of glucose in the cell through action of insulin and cell surface receptors.

2. Covalent Modification of Enzyme- Hexokinase, phosphofructokinase and pyruvate kinase are key enzymes responsible for controlling glycolysis. Most of the typical protein kinases are regulated by a reversible phosphorylation and dephosphorylation. In the presence of low glucose in blood, pyruvate kinase is getting phosphorylated by cytosolic enzymes and phosphoryated pyruvate kinase is less active. Similarly in the presence of high blood glucose level, it remains as unphosphorylated and that relive the inhibition caused by phosphorylation.

Regulation of glycolysis: (A) Covalent Modification (B) Alloteric regulation of enzymes of glycolysis.

3. Allosteric regulation- All the three crucial enzymes Hexokinase, phosphofructokinase and pyruvate kinase of glycolysis are regulated allosterically. In an allosteric regulation, an enzyme binds the allosteric molecules and this modulates the activity of the enzyme either in positive or negative manner. In glycolysis, fructose 2,6 bis phosphate is produced from fructose-6, phosphate by the enzyme phosphofructo kinase-2. fructose 2,6 bis phosphate is allosterically activating the enzymatic activity of phospho fructokinase (PFK-1) and at the same time it is down regulating the activity of fructose 1,6 bis phosphatase. In addition, ATP and citrate is inhibiting the activity of phospho fructokinase where as ADP and AMP is allosterically enhancing the enzymatic activitiy.

Allosteric Regulation

Allosteric regulation of an enzyme

In biochemistry, allosteric regulation (or allosteric control) is the regulation of an enzyme by binding an effector molecule at a site other than the enzyme's active site.

The site to which the effector binds is termed the *allosteric site*. Allosteric sites allow effectors to bind to the protein, often resulting in a conformational change involving protein dynamics. Effectors that enhance the protein's activity are referred to as *allosteric activators*, whereas those that decrease the protein's activity are called *allosteric inhibitors*.

Allosteric regulations are a natural example of control loops, such as feedback from downstream products or feedforward from upstream substrates. Long-range allostery is especially important in cell signaling. Allosteric regulation is also particularly important in the cell's ability to adjust enzyme activity.

The term *allostery* comes from the Greek *allos*, "other," and *stereos*, "solid (object)." This is in reference to the fact that the regulatory site of an allosteric protein is physically distinct from its active site.

Models of Allosteric Regulation

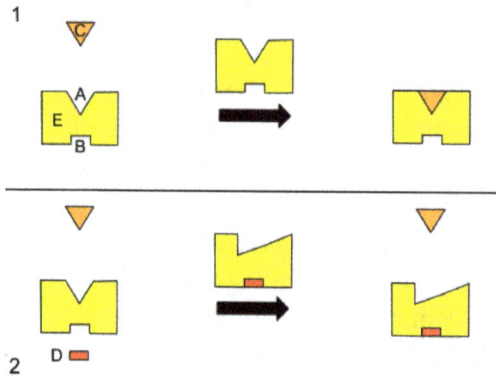

A - Active site B - Allosteric site C - Substrate D - Inhibitor E - Enzyme
This is a diagram of allosteric regulation of an enzyme.

Most allosteric effects can be explained by the *concerted* MWC model put forth by Monod, Wyman, and Changeux, or by the sequential model described by Koshland, Nemethy, and Filmer. Both postulate that enzyme subunits exist in one of two conformations, tensed (T) or relaxed (R), and that relaxed subunits bind substrate more readily than those in the tense state. The two models differ most in their assumptions about subunit interaction and the preexistence of both states.

Concerted Model

The concerted model of allostery, also referred to as the symmetry model or MWC model, postulates that enzyme subunits are connected in such a way that a conformational change in one subunit is necessarily conferred to all other subunits. Thus, all subunits must exist in the same conformation. The model further holds that, in the absence of any ligand (substrate or otherwise), the equilibrium favors one of the conformational states, T or R. The equilibrium can be shifted to the R or T state through the binding of one ligand (the allosteric effector or ligand) to a site that is different from the active site (the allosteric site).

Sequential Model

The sequential model of allosteric regulation holds that subunits are not connected in such a way that a conformational change in one induces a similar change in the others. Thus, all enzyme subunits do not necessitate the same conformation. Moreover, the sequential model dictates that molecules of a substrate bind via an induced fit protocol. In general, when a subunit randomly collides with a molecule of substrate, the active site, in essence, forms a glove around its substrate. While such an induced fit converts a subunit from the tensed state to relaxed state, it does not propagate the conformational change to adjacent subunits. Instead, substrate-binding at one subunit only slightly alters the structure of other subunits so that their binding sites are more receptive to substrate. To summarize:

- subunits need not exist in the same conformation

- molecules of substrate bind via induced-fit protocol

- conformational changes are not propagated to all subunit

Morpheein Model

The morpheein model of allosteric regulation is a dissociative concerted model.

A morpheein is a homo-oligomeric structure that can exist as an ensemble of physiologically significant and functionally different alternate quaternary assemblies. Transitions between alternate morpheein assemblies involve oligomer dissociation, conformational change in the dissociated state, and reassembly to a different oligomer. The required oligomer disassembly step differentiates the morpheein model for allosteric regulation from the classic MWC and KNF models. Porphobilinogen synthase (PBGS) is the prototype morpheein.

Ensemble Models

Ensemble models of allosteric regulation enumerate an allosteric system's statistical ensemble as a function of its potential energy function, and then relate specific statistical measurements of allostery to specific energy terms in the energy function (such as an intermolecular salt bridge between two domains). Ensemble models like the Ensemble Allosteric Model and Allosteric Ising Model assume that each domain of the system can adopt two states similar to the MWC model. The allostery landscape model introduced by Cuendet, Weinstein, and LeVine allows for the domains to have any number of states and the contribution of a specific molecular interaction to a given allosteric coupling can be estimated using a rigorous set of rules. Molecular dynamics simulations can be used to estimate a systems's statistical ensemble so that it can be analyzed with the allostery landscape model.

Allosteric Resources Online

Allosteric Database

Allostery is a direct and efficient means for regulation of biological macromolecule function, produced by the binding of a ligand at an allosteric site topographically distinct from the orthosteric site. Due to the often high receptor selectivity and lower target-based toxicity, allosteric regulation is also expected to play an increasing role in drug discovery and bioengineering. The AlloSteric Database provides a central resource for the display, search and analysis of the structure, function and related annotation for allosteric molecules. Currently, ASD contains allosteric proteins from more than 100 species and modulators in three categories (activators, inhibitors, and regulators). Each protein is annotated with detailed description of allostery, biological process and related diseases, and each modulator with binding affinity, physicochemical properties and therapeutic area. Integrating the information of allosteric proteins in ASD should allow the prediction of allostery for unknown proteins, to be followed with experimental validation. In addition, modulators curated in ASD can be used to investigate potential allosteric targets for a query compound, and can help chemists to implement structure modifications for novel allosteric drug design.

Allosteric Residues and their Prediction using the Stress Web Server

Not all protein residues play equally important roles in allosteric regulation. The identification of residues that are essential to allostery (so-called "allosteric residues") has been the focus of many studies, especially within the last decade. In part, this growing interest is a result of their general importance in protein science, but also because allosteric residues may be exploited in biomedical contexts. Pharmacologically important proteins with difficult-to-target sites may yield to approaches in which one alternatively targets easier-to-reach residues that are capable of allosterically regulating the primary site of interest. These residues can broadly be classified as surface- and interior-allosteric amino acids. Allosteric sites at the surface generally play regulatory roles that are fundamentally distinct from those within the interior; surface residues may serve as receptors or effector sites in allosteric signal transmission, whereas those within the interior may act to transmit such signals. STRucturally-identified ESSential residues is a web tool that enables users to submit their own protein structures of interest in order to predict both surface- and interior-allosteric residues in an algorithmically efficient manner. The software behind this server employs 3D structures to build models of conformational change in order to perform predictions.

Allosteric Modulation

Positive Modulation

Positive allosteric modulation (also known as *allosteric activation*) occurs when the binding of one ligand enhances the attraction between substrate molecules and other binding sites. An example is the binding of oxygen molecules to hemoglobin, where oxygen is effectively both the substrate and the effector. The allosteric, or "other", site is the active site of an adjoining protein subunit. The binding of oxygen to one subunit induces a conformational change in that subunit that interacts with the remaining active sites to enhance *their* oxygen affinity. Another example of allosteric activation is seen in cytosolic IMP-GMP specific 5'-nucleotidase II (cN-II), where the affinity for substrate GMP increases upon GTP binding at the dimer interface

Negative Modulation

Negative allosteric modulation (also known as *allosteric inhibition*) occurs when the binding of one ligand decreases the affinity for substrate at other active sites. For example, when 2,3-BPG binds to an allosteric site on hemoglobin, the affinity for oxygen of all subunits decreases. This is when a regulator is absent from the binding site.

Direct thrombin inhibitors provides an excellent example of negative allosteric modulation. Allosteric inhibitors of thrombin have been discovered which could potentially be used as anticoagulants.

Another example is strychnine, a convulsant poison, which acts as an allosteric inhibitor of the glycine receptor. Glycine is a major post-synaptic inhibitory neurotransmitter in mammalian spinal cord and brain stem. Strychnine acts at a separate binding site on the glycine receptor in an allosteric manner; i.e., its binding lowers the affinity of the glycine receptor for glycine. Thus, strychnine inhibits the action of an inhibitory transmitter, leading to convulsions.

Another instance in which negative allosteric modulation can be seen is between ATP and the enzyme Phosphofructokinase within the negative feedback loop that regulates glycolysis. Phosphofructokinase (generally referred to as PFK) is an enzyme that catalyses the third step of glycolysis: the phosphorylation of Fructose-6-phosphate into Fructose 1,6-bisphosphate. PFK can be allosterically inhibited by high levels of ATP within the cell. When ATP levels are high, ATP will bind to an allosteoric site on phosphofructokinase, causing a change in the enzyme's three-dimensional shape. This change causes its affinity for substrate (fructose-6-phosphate and ATP) at the active site to decrease, and the enzyme is deemed inactive. This causes glycolysis to cease when ATP levels are high, thus conserving the body's glucose and maintaining balanced levels of cellular ATP. In this way, ATP serves as a negative allosteric modulator for PFK, despite the fact that it is also a substrate of the enzyme.

Types

Homotropic

A homotropic allosteric modulator is a substrate for its target enzyme, as well as a regulatory molecule of the enzyme's activity. It is typically an activator of the enzyme. For example, O_2 and CO_2 are homotropic allosteric modulators of hemoglobin.

Heterotropic

A heterotropic allosteric modulator is a regulatory molecule that is not the enzyme's substrate. It may be either an activator or an inhibitor of the enzyme. For example, H^+, CO_2, and 2,3-bisphosphoglycerate are heterotropic allosteric modulators of hemoglobin.

Some allosteric proteins can be regulated by both their substrates and other molecules. Such proteins are capable of both homotropic and heterotropic interactions.

Non-regulatory Allostery

A non-regulatory allosteric site refers to any non-regulatory component of an enzyme (or any protein), that is not itself an amino acid. For instance, many enzymes require sodium binding to ensure proper function. However, the sodium does not necessarily act as a regulatory subunit; the sodium is always present and there are no known biological processes to add/remove sodium to regulate enzyme activity. Non-regulatory allostery could comprise any other ions besides sodium (calcium, magnesium, zinc), as well as other chemicals and possibly vitamins.

Pharmacology

Allosteric modulation of a receptor results from the binding of allosteric modulators at a different site (a "regulatory site") from that of the endogenous ligand (an "active site") and enhances or inhibits the effects of the endogenous ligand. Under normal circumstances, it acts by causing a conformational change in a receptor molecule, which results in a change in the binding affinity of the ligand. In this way, an allosteric ligand modulates the receptor's activation by its primary (orthosteric) ligand, and can be thought to act like a dimmer switch in an electrical circuit, adjusting the intensity of the response.

For example, the GABA$_A$ receptor has two active sites that the neurotransmitter gamma-aminobutyric acid (GABA) binds, but also has benzodiazepine and general anaesthetic agent regulatory binding sites. These regulatory sites can each produce positive allosteric modulation, potentiating the activity of GABA. Diazepam is an agonist at the benzodiazepine regulatory site, and its antidote flumazenil is an antagonist.

More recent examples of drugs that allosterically modulate their targets include the calcium-mimicking cinacalcet and the HIV treatment maraviroc.

Allosteric Sites as Drug Targets

Allosteric sites may represent a novel drug target. There are a number of advantages in using allosteric modulators as preferred therapeutic agents over classic orthosteric ligands. For example, G protein-coupled receptor (GPCR) allosteric binding sites have not faced the same evolutionary pressure as orthosteric sites to accommodate an endogenous ligand, so are more diverse. Therefore, greater GPCR selectivity may be obtained by targeting allosteric sites. This is particularly useful for GPCRs where selective orthosteric therapy has been difficult because of sequence conservation of the orthosteric site across receptor subtypes. Also, these modulators have a decreased potential for toxic effects, since modulators with limited co-operativity will have a ceiling level to their effect, irrespective of the administered dose. Another type of pharmacological selectivity that is unique to allosteric modulators is based on co-operativity. An allosteric modulator may display neutral co-operativity with an orthosteric ligand at all subtypes of a given receptor except the subtype of interest, which is termed "absolute subtype selectivity". If an allosteric modulator does not possess appreciable efficacy, it can provide another powerful therapeutic advantage over orthosteric ligands, namely the ability to selectively tune up or down tissue responses only when the endogenous agonist is present. Oligomer-specific small molecule binding sites are drug targets for medically relevant morpheeins.

Citric Acid Cycle

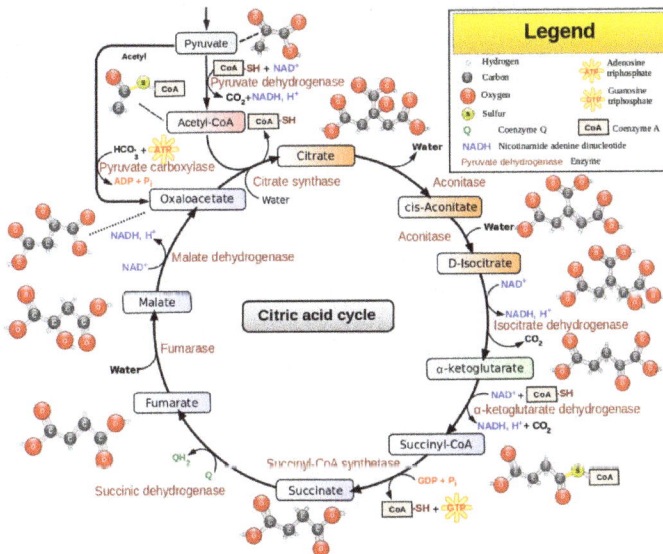

Overview of the citric acid cycle

The citric acid cycle (CAC) – also known as the tricarboxylic acid (TCA) cycle or the Krebs cycle – is a series of chemical reactions used by all aerobic organisms to release stored energy through the oxidation of acetyl-CoA derived from carbohydrates, fats and proteins into carbon dioxide and chemical energy in the form of adenosine triphosphate, (ATP.) In addition, the cycle provides precursors of certain amino acids as well as the reducing agent NADH that is used in numerous other biochemical reactions. Its central importance to many biochemical pathways suggests that it was one of the earliest established components of cellular metabolism and may have originated abiogenically.

The name of this metabolic pathway is derived from citric acid (a type of tricarboxylic acid, often called citrate, as the ionized form predominates at biological pH) that is consumed and then regenerated by this sequence of reactions to complete the cycle. In addition, the cycle consumes acetate (in the form of acetyl-CoA) and water, reduces NAD^+ to NADH, and produces carbon dioxide as a waste byproduct. The NADH generated by the TCA cycle is fed into the oxidative phosphorylation (electron transport) pathway. The net result of these two closely linked pathways is the oxidation of nutrients to produce usable chemical energy in the form of ATP.

In eukaryotic cells, the citric acid cycle occurs in the matrix of the mitochondrion. In prokaryotic cells, such as bacteria which lack mitochondria, the TCA reaction sequence is performed in the cytosol with the proton gradient for ATP production being across the cell's surface (plasma membrane) rather than the inner membrane of the mitochondrion.

Discovery

Several of the components and reactions of the citric acid cycle were established in the 1930s by the research of the Albert Szent-Györgyi, who received the Nobel Prize in Physiology or Medicine in 1937 specifically for his discoveries pertaining to fumaric acid; a key component of the cycle. The citric acid cycle itself was finally identified in 1937 by Hans Adolf Krebs and William Arthur Johnson while at the University of Sheffield, for which the former received the Nobel Prize for Physiology or Medicine in 1953.

Evolution

Components of the TCA cycle were derived from anaerobic bacteria, and the TCA cycle itself may have evolved more than once. Theoretically there are several alternatives to the TCA cycle; however, the TCA cycle appears to be the most efficient. If several TCA alternatives had evolved independently, they all appear to have converged to the TCA cycle.

Overview

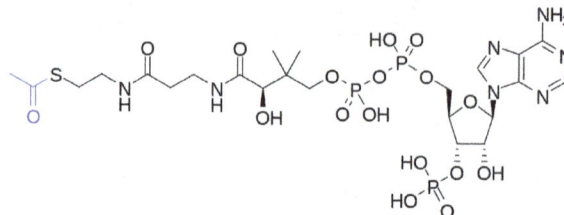

Structural diagram of acetyl-CoA. The portion in blue, on the left, is the acetyl group; the portion in black is coenzyme A.

The citric acid cycle is a key metabolic pathway that connects carbohydrate, fat, and protein metabolism. The reactions of the cycle are carried out by 8 enzymes that completely oxidize acetate, in the form of acetyl-CoA, into two molecules each of carbon dioxide and water. Through catabolism of sugars, fats, and proteins, the two-carbon organic product acetyl-CoA (a form of acetate) is produced which enters the citric acid cycle. The reactions of the cycle also convert three equivalents of nicotinamide adenine dinucleotide (NAD⁺) into three equivalents of reduced NAD⁺ (NADH), one equivalent of flavin adenine dinucleotide (FAD) into one equivalent of $FADH_2$, and one equivalent each of guanosine diphosphate (GDP) and inorganic phosphate (P_i) into one equivalent of guanosine triphosphate (GTP). The NADH and $FADH_2$ generated by the citric acid cycle are in turn used by the oxidative phosphorylation pathway to generate energy-rich adenosine triphosphate (ATP).

One of the primary sources of acetyl-CoA is from the breakdown of sugars by glycolysis which yield pyruvate that in turn is decarboxylated by the enzyme pyruvate dehydrogenase generating acetyl-CoA according to the following reaction scheme:

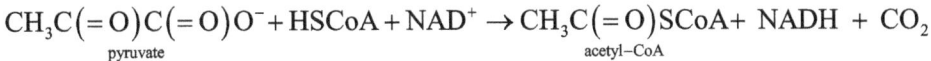

$$CH_3C(=O)C(=O)O^- + HSCoA + NAD^+ \rightarrow CH_3C(=O)SCoA + NADH + CO_2$$

pyruvate   acetyl–CoA

The product of this reaction, acetyl-CoA, is the starting point for the citric acid cycle. Acetyl-CoA may also be obtained from the oxidation of fatty acids. Below is a schematic outline of the cycle:

- The citric acid cycle begins with the transfer of a two-carbon acetyl group from acetyl-CoA to the four-carbon acceptor compound (oxaloacetate) to form a six-carbon compound (citrate).

- The citrate then goes through a series of chemical transformations, losing two carboxyl groups as CO_2. The carbons lost as CO_2 originate from what was oxaloacetate, not directly from acetyl-CoA. The carbons donated by acetyl-CoA become part of the oxaloacetate carbon backbone after the first turn of the citric acid cycle. Loss of the acetyl-CoA-donated carbons as CO_2 requires several turns of the citric acid cycle. However, because of the role of the citric acid cycle in anabolism, they might not be lost, since many TCA cycle intermediates arc also used as precursors for the biosynthesis of other molecules.

- Most of the energy made available by the oxidative steps of the cycle is transferred as energy-rich electrons to NAD⁺, forming NADH. For each acetyl group that enters the citric acid cycle, three molecules of NADH are produced.

- Electrons are also transferred to the electron acceptor Q, forming QH_2 (Q = FAD+, QH_2 = $FADH_2$).

- For every NADH and $FADH_2$ that are produced in the citric acid cycle, 2.5 and 1.5 ATP molecules are generated in oxidative phosphorylation, respectively.

- At the end of each cycle, the four-carbon oxaloacetate has been regenerated, and the cycle continues.

Steps

Two carbon atoms are oxidized to CO_2, the energy from these reactions is transferred to other metabolic processes through GTP (or ATP), and as electrons in NADH and QH_2. The NADH generated in the TCA cycle may later be oxidized (donate its electrons) to drive ATP synthesis in a type of process called oxidative phosphorylation. $FADH_2$ is covalently attached to succinate dehydrogenase, an enzyme which functions both in the CAC and the mitochondrial electron transport chain in oxidative phosphorylation. $FADH_2$, therefore, facilitates transfer of electrons to coenzyme Q, which is the final electron acceptor of the reaction catalyzed by the Succinate:ubiquinone oxidoreductase complex, also acting as an intermediate in the electron transport chain.

The citric acid cycle is continuously supplied with new carbon in the form of acetyl-CoA, entering at step 0 below.

	Substrates	Products	Enzyme	Reaction type	Comment
0 / 10	Oxaloacetate + Acetyl CoA + H_2O	Citrate + CoA-SH	Citrate synthase	Aldol condensation	irreversible, extends the 4C oxaloacetate to a 6C molecule
1	Citrate	*cis*-Aconitate + H_2O	Aconitase	Dehydration	reversible isomerisation
2	*cis*-Aconitate + H_2O	Isocitrate		Hydration	
3	Isocitrate + NAD^+	Oxalosuccinate + NADH + H^+	Isocitrate dehydrogenase	Oxidation	generates NADH (equivalent of 2.5 ATP)
4	Oxalosuccinate	α-Ketoglutarate + CO_2		Decarboxylation	rate-limiting, irreversible stage, generates a 5C molecule
5	α-Ketoglutarate + NAD^+ + CoA-SH	Succinyl-CoA + NADH + H^+ + CO_2	α-Ketoglutarate dehydrogenase	Oxidative decarboxylation	irreversible stage, generates NADH (equivalent of 2.5 ATP), regenerates the 4C chain (CoA excluded)
6	Succinyl-CoA + GDP + P_i	Succinate + CoA-SH + GTP	Succinyl-CoA synthetase	substrate-level phosphorylation	or ADP→ATP instead of GDP→GTP, generates 1 ATP or equivalent Condensation reaction of GDP + P_i and hydrolysis of Succinyl-CoA involve the H_2O needed for balanced equation.

7	Succinate + ubiquinone (Q)	Fumarate + ubiquinol (QH$_2$)	Succinate dehy-drogenase	Oxidation	uses FAD as a prosthetic group (FAD→FADH$_2$ in the first step of the reaction) in the enzyme, generates the equivalent of 1.5 ATP
8	Fumarate + H$_2$O	L-Malate	Fumarase	Hydration	Hydration of C-C double bond
9	L-Malate + NAD$^+$	Oxaloace-tate + NADH + H$^+$	Malate dehydro-genase	Oxidation	reversible (in fact, equilibri-um favors malate), generates NADH (equivalent of 2.5 ATP)
10 / 0	Oxaloacetate + Acetyl CoA + H$_2$O	Citrate + CoA-SH	Citrate synthase	Aldol conden-sation	This is the same as step 0 and restarts the cycle. The reaction is irreversible and extends the 4C oxaloacetate to a 6C molecule

Mitochondria in animals, including humans, possess two succinyl-CoA synthetases: one that produces GTP from GDP, and another that produces ATP from ADP. Plants have the type that produces ATP (ADP-forming succinyl-CoA synthetase). Several of the enzymes in the cycle may be loosely associated in a multienzyme protein complex within the mitochondrial matrix.

The GTP that is formed by GDP-forming succinyl-CoA synthetase may be utilized by nucleoside-diphosphate kinase to form ATP (the catalyzed reaction is GTP + ADP → GDP + ATP).

Products

Products of the first turn of the cycle are: *one GTP (or ATP), three NADH, one QH$_2$, two CO$_2$.*

Because two acetyl-CoA molecules are produced from each glucose molecule, two cycles are required per glucose molecule. Therefore, at the end of two cycles, the products are: two GTP, six NADH, two QH$_2$, and four CO$_2$

Description	Reactants	Products
The sum of all reactions in the citric acid cycle is:	Acetyl-CoA + 3 NAD$^+$ + Q + GDP + P$_i$ + 2 H$_2$O	→ CoA-SH + 3 NADH + 3 H$^+$ + QH$_2$ + GTP + 2 CO$_2$
Combining the reactions occurring during the pyru-vate oxidation with those occurring during the citric acid cycle, the following overall pyruvate oxidation re-action is obtained:	Pyruvate ion + 4 NAD$^+$ + Q + GDP + P$_i$ + 2 H$_2$O	→ 4 NADH + 4 H$^+$ + QH$_2$ + GTP + 3 CO$_2$

Combining the above reaction with the ones occurring in the course of glycolysis, the following overall glucose oxidation reaction (excluding reactions in the respiratory chain) is obtained:	Glucose + 10 NAD$^+$ + 2 Q + 2 ADP + 2 GDP + 4 P$_i$ + 2 H$_2$O	$\rightarrow$ 10 NADH + 10 H$^+$ + 2 QH$_2$ + 2 ATP + 2 GTP + 6 CO$_2$

The above reactions are balanced if P$_i$ represents the H$_2$PO$_4^-$ ion, ADP and GDP the ADP^{2-} and GDP^{2-} ions, respectively, and ATP and GTP the ATP^{3-} and GTP^{3-} ions, respectively.

The total number of ATP molecules obtained after complete oxidation of one glucose in glycolysis, citric acid cycle, and oxidative phosphorylation is estimated to be between 30 and 38.

Efficiency

The theoretical maximum yield of ATP through oxidation of one molecule of glucose in glycolysis, citric acid cycle, and oxidative phosphorylation is 38 (assuming 3 molar equivalents of ATP per equivalent NADH and 2 ATP per FADH$_2$). In eukaryotes, two equivalents of NADH are generated in glycolysis, which takes place in the cytoplasm. Transport of these two equivalents into the mitochondria consumes two equivalents of ATP, thus reducing the net production of ATP to 36. Furthermore, inefficiencies in oxidative phosphorylation due to leakage of protons across the mitochondrial membrane and slippage of the ATP synthase/proton pump commonly reduces the ATP yield from NADH and FADH$_2$ to less than the theoretical maximum yield. The observed yields are, therefore, closer to ~2.5 ATP per NADH and ~1.5 ATP per FADH$_2$, further reducing the total net production of ATP to approximately 30. An assessment of the total ATP yield with newly revised proton-to-ATP ratios provides an estimate of 29.85 ATP per glucose molecule.

Variation

While the TCA cycle is in general highly conserved, there is significant variability in the enzymes found in different taxa (note that the diagrams on this page are specific to the mammalian pathway variant).

Some differences exist between eukaryotes and prokaryotes. The conversion of D-*threo*-isocitrate to 2-oxoglutarate is catalyzed in eukaryotes by the NAD$^+$-dependent EC 1.1.1.41, while prokaryotes employ the NADP$^+$-dependent EC 1.1.1.42. Similarly, the conversion of (S)-malate to oxaloacetate is catalyzed in eukaryotes by the NAD$^+$-dependent EC 1.1.1.37, while most prokaryotes utilize a quinone-dependent enzyme, EC 1.1.5.4.

A step with significant variability is the conversion of succinyl-CoA to succinate. Most organisms utilize EC 6.2.1.5, succinate–CoA ligase (ADP-forming) (despite its name, the enzyme operates in the pathway in the direction of ATP formation). In mammals a

GTP-forming enzyme, succinate–CoA ligase (GDP-forming) (EC 6.2.1.4) also operates. The level of utilization of each isoform is tissue dependent. In some acetate-producing bacteria, such as *Acetobacter aceti*, an entirely different enzyme catalyzes this conversion – EC 2.8.3.18, succinyl-CoA:acetate CoA-transferase. This specialized enzyme links the TCA cycle with acetate metabolism in these organisms. Some bacteria, such as *Helicobacter pylori*, employ yet another enzyme for this conversion – succinyl-CoA:acetoacetate CoA-transferase (EC 2.8.3.5).

Some variability also exists at the previous step – the conversion of 2-oxoglutarate to succinyl-CoA. While most organisms utilize the ubiquitous NAD^+-dependent 2-oxoglutarate dehydrogenase, some bacteria utilize a ferredoxin-dependent 2-oxoglutarate synthase (EC 1.2.7.3). Other organisms, including obligately autotrophic and methanotrophic bacteria and archaea, bypass succinyl-CoA entirely, and convert 2-oxoglutarate to succinate via succinate semialdehyde, using EC 4.1.1.71, 2-oxoglutarate decarboxylase, and EC 1.2.1.79, succinate-semialdehyde dehydrogenase.

Regulation

The regulation of the TCA cycle is largely determined by product inhibition and substrate availability. If the cycle were permitted to run unchecked, large amounts of metabolic energy could be wasted in overproduction of reduced coenzyme such as NADH and ATP. The major eventual substrate of the cycle is ADP which gets converted to ATP. A reduced amount of ADP causes accumulation of precursor NADH which in turn can inhibit a number of enzymes. NADH, a product of all dehydrogenases in the TCA cycle with the exception of succinate dehydrogenase, inhibits pyruvate dehydrogenase, isocitrate dehydrogenase, α-ketoglutarate dehydrogenase, and also citrate synthase. Acetyl-coA inhibits pyruvate dehydrogenase, while succinyl-CoA inhibits alpha-ketoglutarate dehydrogenase and citrate synthase. When tested in vitro with TCA enzymes, ATP inhibits citrate synthase and α-ketoglutarate dehydrogenase; however, ATP levels do not change more than 10% in vivo between rest and vigorous exercise. There is no known allosteric mechanism that can account for large changes in reaction rate from an allosteric effector whose concentration changes less than 10%.

Calcium is also used as a regulator in the TCA cycle. Calcium levels in the mitochondrial matrix can reach up to the tens of micromolar levels during cellular activation. It activates pyruvate dehydrogenase phosphatase which in turn activates the pyruvate dehydrogenase complex. Calcium also activates isocitrate dehydrogenase and α-ketoglutarate dehydrogenase. This increases the reaction rate of many of the steps in the cycle, and therefore increases flux throughout the pathway.

Citrate is used for feedback inhibition, as it inhibits phosphofructokinase, an enzyme involved in glycolysis that catalyses formation of fructose 1,6-bisphosphate, a precursor of pyruvate. This prevents a constant high rate of flux when there is an accumulation of citrate and a decrease in substrate for the enzyme.

Recent work has demonstrated an important link between intermediates of the citric acid cycle and the regulation of hypoxia-inducible factors (HIF). HIF plays a role in the regulation of oxygen homeostasis, and is a transcription factor that targets angio-genesis, vascular remodeling, glucose utilization, iron transport and apoptosis. HIF is synthesized consititutively, and hydroxylation of at least one of two critical proline residues mediates their interaction with the von Hippel Lindau E3 ubiquitin ligase complex, which targets them for rapid degradation. This reaction is catalysed by prolyl 4-hydroxylases. Fumarate and succinate have been identified as potent inhibitors of prolyl hydroxylases, thus leading to the stabilisation of HIF.

Major Metabolic Pathways Converging on the TCA Cycle

Several catabolic pathways converge on the TCA cycle. Most of these reactions add intermediates to the TCA cycle, and are therefore known as anaplerotic reactions. These increase the amount of acetyl CoA that the cycle is able to carry, increasing the mitochondrion's capability to carry out respiration if this is otherwise a limiting factor. Processes that remove intermediates from the cycle are termed "cataplerotic" reactions.

Pyruvate molecules produced by glycolysis are actively transported across the inner mi-tochondrial membrane, and into the matrix. Here they can be oxidized and combined with coenzyme A to form CO_2, *acetyl-CoA*, and NADH, as in the normal cycle.

However, it is also possible for pyruvate to be carboxylated by pyruvate carboxylase to form *oxaloacetate*. This latter reaction "fills up" the amount of *oxaloacetate* in the citric acid cycle, and is therefore an anaplerotic reaction, increasing the cycle's capacity to metabolize *acetyl-CoA* when the tissue's energy needs (e.g. in muscle) are suddenly increased by activity.

In the citric acid cycle all the intermediates (e.g. *citrate, iso-citrate, alpha-ketogluta-rate, succinate, fumarate, malate* and *oxaloacetate*) are regenerated during each turn of the cycle. Adding more of any of these intermediates to the mitochondrion therefore means that that additional amount is retained within the cycle, increasing all the other intermediates as one is converted into the other. Hence the addition of any one of them to the cycle has an anaplerotic effect, and its removal has a cataplerotic effect. These anaplerotic and cataplerotic reactions will, during the course of the cycle, increase or decrease the amount of *oxaloacetate* available to combine with *acetyl-CoA* to form *cit-ric acid*. This in turn increases or decreases the rate of ATP production by the mito-chondrion, and thus the availability of ATP to the cell.

Acetyl-CoA, on the other hand, derived from pyruvate oxidation, or from the beta-ox-idation of fatty acids, is the only fuel to enter the citric acid cycle. With each turn of the cycle one molecule of *acetyl-CoA* is consumed for every molecule of *oxaloacetate* present in the mitochondrial matrix, and is never regenerated. It is the oxidation of the acetate portion of *acetyl-CoA* that produces CO_2 and water, with the energy thus released captured in the form of ATP.

In the liver, the carboxylation of cytosolic pyruvate into intra-mitochondrial *oxaloacetate* is an early step in the gluconeogenic pathway which converts lactate and de-aminated alanine into glucose, under the influence of high levels of glucagon and/or epinephrine in the blood. Here the addition of *oxaloacetate* to the mitochondrion does not have a net anaplerotic effect, as another citric acid cycle intermediate (*malate*) is immediately removed from the mitochondrion to be converted into cytosolic oxaloacetate, which is ultimately converted into glucose, in a process that is almost the reverse of glycolysis.

In protein catabolism, proteins are broken down by proteases into their constituent amino acids. Their carbon skeletons (i.e. the de-aminated amino acids) may either enter the citric acid cycle as intermediates (e.g. *alpha-ketoglutarate* derived from glutamate or glutamine), having an anaplerotic effect on the cycle, or, in the case of leucine, isoleucine, lysine, phenylalanine, tryptophan, and tyrosine, they are converted into *acetyl-CoA* which can be burned to CO_2 and water, or used to form ketone bodies, which too can only be burned in tissues other than the liver where they are formed, or excreted via the urine or breath. These latter amino acids are therefore termed "ketogenic" amino acids, whereas those that enter the citric acid cycle as intermediates can only be cataplerotically removed by entering the gluconeogenic pathway via *malate* which is transported out of the mitochondrion to be converted into cytosolic oxaloacetate and ultimately into glucose. These are the so-called "glucogenic" amino acids. De-aminated alanine, cysteine, glycine, serine, and threonine are converted to pyruvate and can consequently either enter the citric acid cycle as *oxaloacetate* (an anaplerotic reaction) or as *acetyl-CoA* to be disposed of as CO_2 and water.

In fat catabolism, triglycerides are hydrolyzed to break them into fatty acids and glycerol. In the liver the glycerol can be converted into glucose via dihydroxyacetone phosphate and glyceraldehyde-3-phosphate by way of gluconeogenesis. In many tissues, especially heart and skeletal muscle tissue, fatty acids are broken down through a process known as beta oxidation, which results in the production of mitochondrial *acetyl-CoA*, which can be used in the citric acid cycle. Beta oxidation of fatty acids with an odd number of methylene bridges produces propionyl-CoA, which is then converted into *succinyl-CoA* and fed into the citric acid cycle as an anaplerotic intermediate.

The total energy gained from the complete breakdown of one (six-carbon) molecule of glucose by glycolysis, the formation of 2 *acetyl-CoA* molecules, their catabolism in the citric acid cycle, and oxidative phosphorylation equals about 30 ATP molecules, in eukaryotes. The number of ATP molecules derived from the beta oxidation of a 6 carbon segment of a fatty acid chain, and the subsequent oxidation of the resulting 3 molecules of *acetyl-CoA* is 40.

Citric Acid Cycle Intermediates Serve as Substrates for Biosynthetic Processes

Several of the citric acid cycle intermediates are used for the synthesis of important compounds, which will have significant cataplerotic effects on the cycle. *Acetyl-CoA* cannot be transported out of the mitochondrion. To obtain cytosolic acetyl-CoA, *citrate* is removed from the citric acid cycle and carried across the inner mitochondrial membrane into the cytosol. There it is cleaved by ATP citrate lyase into acetyl-CoA and oxaloacetate. The oxaloacetate is returned to mitochondrion as *malate* (and then converted back into *oxaloacetate* to transfer more *acetyl-CoA* out of the mitochondrion). The cytosolic acetyl-CoA is used for fatty acid synthesis and the production of cholesterol. Cholesterol can, in turn, be used to synthesize the steroid hormones, bile salts, and vitamin D.

The carbon skeletons of many non-essential amino acids are made from citric acid cycle intermediates. To turn them into amino acids the alpha keto-acids formed from the citric acid cycle intermediates have to acquire their amino groups from glutamate in a transamination reaction, in which pyridoxal phosphate is a cofactor. In this reaction the glutamate is converted into *alpha-ketoglutarate*, which is a citric acid cycle intermediate. The intermediates that can provide the carbon skeletons for amino acid synthesis are *oxaloacetate* which forms aspartate and asparagine; and *alpha-ketoglutarate* which forms glutamine, proline, and arginine.

Of these amino acids, aspartate and glutamine are used, together with carbon and nitrogen atoms from other sources, to form the purines that are used as the bases in DNA and RNA, as well as in ATP, AMP, GTP, NAD, FAD and CoA.

The pyrimidines are partly assembled from aspartate (derived from *oxaloacetate*). The pyrimidines, thymine, cytosine and uracil, form the complementary bases to the purine bases in DNA and RNA, and are also components of CTP, UMP, UDP and UTP.

The majority of the carbon atoms in the porphyrins come from the citric acid cycle intermediate, *succinyl-CoA*. These molecules are an important component of the hemoproteins, such as hemoglobin, myoglobin and various cytochromes.

During gluconeogenesis mitochondrial *oxaloacetate* is reduced to *malate* which is then transported out of the mitochondrion, to be oxidized back to oxaloacetate in the cytosol. Cytosolic oxaloacetate is then decarboxylated to phosphoenolpyruvate by phosphoenolpyruvate carboxykinase, which is the rate limiting step in the conversion of nearly all the gluconeogenic precursors (such as the glucogenic amino acids and lactate) into glucose by the liver and kidney.

Because the citric acid cycle is involved in both catabolic and anabolic processes, it is known as an amphibolic pathway.

Interactive Pathway Map

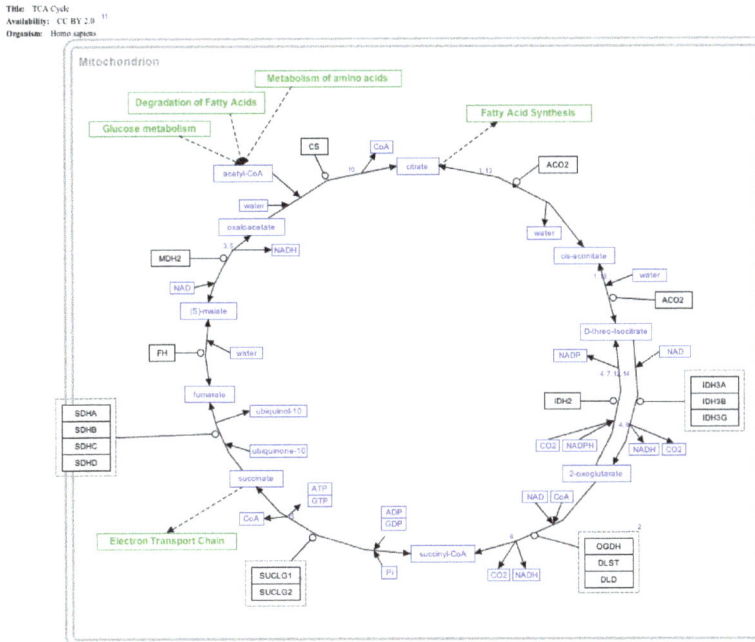

Anaerobic Oxidation

Anaerobic Oxidation- Glucose enters into the glycolysis produce pyruvate, which in turn enters into the Kreb's cycle for complete oxidation to produce maximum energy. The primary requirement of the oxidative phosphorylation is presence of a well developed electron transport chain to process reducing equivalents to produce ATP. In addition, presence of oxygen is mandatory for this process. Hence, depending upon the environmental conditions, pyruvate produced in glycolysis has multiple routes to follow as given in the figure below. As discussed before, in the presence of oxygen, pyruvate directly enters into the kreb cycle to follow oxidative phosphorylation. In the absence of oxygen, pyruvate accumulates in cytosol and is immediately processed into two routes: (1) direct conversion to lactate with the help of cytosolic enzyme lactate dehyrogenase (LDH). (2) conversion of pyruvate to alchol with acetaldehyde as a intermediate by the concerted action of pyruvate decarboxylase and alchol dehyrogenase.

These set of reactions operating in the absence of oxygen helps organism in many ways. Now we will discuss the mechanism of pyruvate conversion to lactic acid or alchol and the significance of these pathways in adopting to the low oxygen environment.

Anaerobic reduction of Pyruvate to Lactate- Pyruvate is reduced to lactate with an enzymatic action of lactate dehyrogenase. In this process, cell spend 1 molecule of NADH and 1 molecule of NAD^+ is generated. The NAD^+ produced in this process will be used to continue running glycolysis and other metabolic pathways.

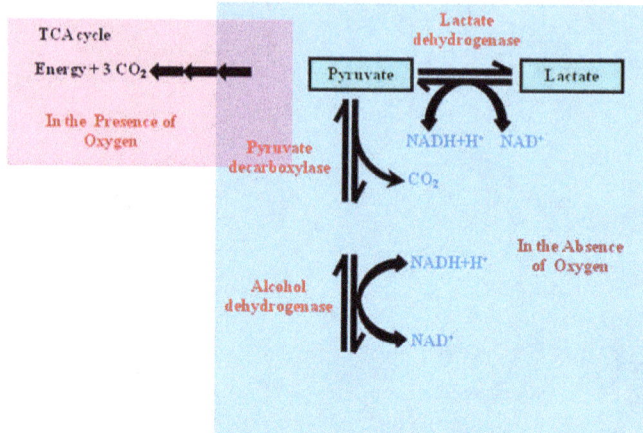

The distribution of Pyruvate during carbohydrate metabolism.

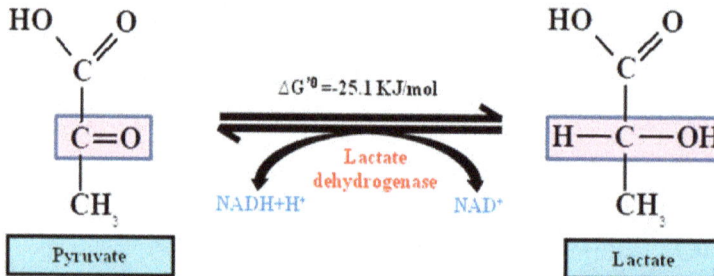

Conversion of Pyruvate to Lactate.

The free energy change (-25.1 KJ/mol) of the pyruvate to lactate conversion favors lactate formation inspite of no net gain of NADH, but it allows the glycolysis to keep running in the absence of oxygen.

Pyruvate to Ethanol- It is a two step process, first conversion of pyruvate to acetaldehyde and in the second step conversion of acetaldehyde to alchol. First step is a decarboxylation reaction catalyzed by pyruvate decarboxylase where as second step is reduction reaction catalyzed by alchol dehydrogenase.

The Over all equation of ethanol production from pyruvate is as follows

$$\text{Glucose} + 2ADP + 2Pi \rightarrow 2 \text{ ethanol} + 2CO_2 + 2ATP + 2H_2O$$

Mechanism of pyruvate decarboxylation by pyruvate decarboxylase- Pyruvate decarboxylase requires thiamine pyrophosphate (TPP) and Mg^{2+} as cofactors to catalyze decarboxylation reaction. Thiamine pyrophosphate is a co-enzyme present in pyruvate decarboxylase and responsible for stabilizing carbanion intermediate. The sequence of event of reaction catalyzed by pyruvate decarboxylase is as follows-

1. Deprotination of TPP to form TPP carbanion.

2. Carbanion attacks on carbonyl group of pyruvate to form adduct.

3. Release of CO_2 .

4. Resonance stabilization of intermediates.

5. Protonation to generate hydroxyl methyl TPP.

6. Release of acetaldehyde and regeneration of TPP from hydroxyl-TPP for next round of enzymatic catalysis.

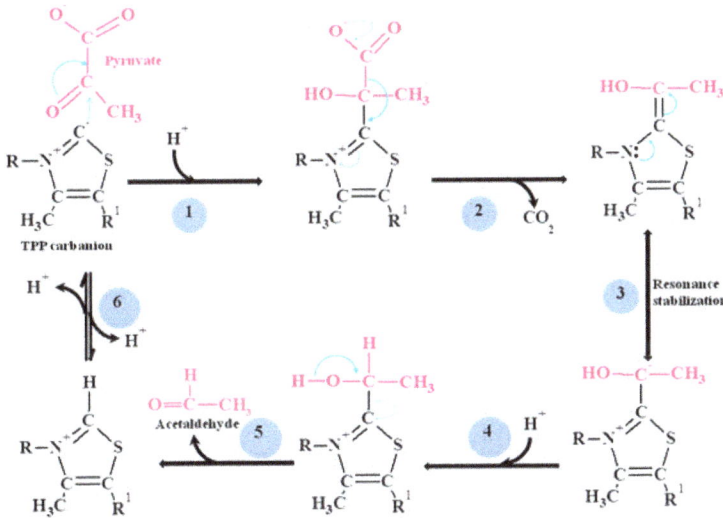

Mechanistic details of conversion of pyruvate to acetaldehyde by pyruvate decarboxylase.

Mechanism of Acetaldehyde to alchol- Alchol dehydrogenase is a dimeric metal dependent dehydrogenase present in animal, plant and bacteria. The reaction mechanism discussed below might have some modifications but over-all alcohol dehyrogenase follows it. The conversion of acetaldehyde to alcohol by alcohol dehydrogenase completes in 4 steps:

1. Binding of substrate acetaldehyde to enzyme bound zinc,

2. Binding of NADH

3. Transfer of hydride ion from NADH to reduce acetaldehyde.

4. Reduced acetaldehyde intermediate acquires a proton from water to form alcohol.

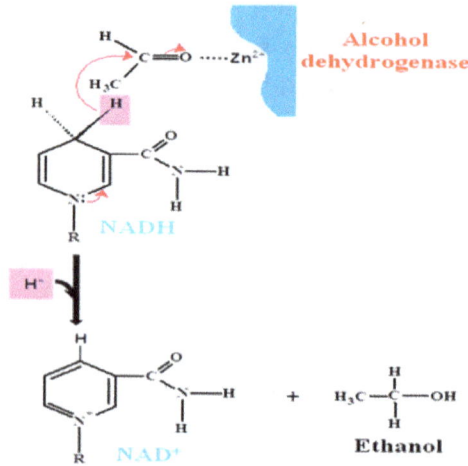

Mechanistic details of conversion of acetaldehyde to alchol by alchol dehydrogenase

The balance sheet of suring fermentation of Glucose to Alcohol

The balance sheet is as follow- STEPS OF GLY-COLYSI	Number of ATP Generation (+) of Investment (-)
1. Step 1-4	-2
2. Generation of 2 molecules of glyceraldehyde-3 phosphate.	
3. Step 6, generation of NADH, Each NADH in ETS gives 3 ATP	3×2=6 [If ETS will operate]
4. Step 7, Generation of ATP	2×1=2
5. Step 10, Generation of ATP	2×1=2
6. Oxidation of one glucose molecule.	6+2+2+2=8
7. Pyruvate to acetaldehyde	0
8. Acetaldehyde to alchol, NADH	-3×2=6
9. NET BALANCE	8-6=2 ATP molecule per glucose

Significance of Anaerobic Oxidation

In the absence of oxygen, cell becomes short of NAD^+ as glycolysis convert all NAD^+ into the NADH. Kreb's cycle is not operating and to continue glycolysis to produce energy, NAD^+ is required. To meet the requirement of maintaining NAD^+ pool, metabolism has adopted a futile cycle approach where NADH produced in glycolysis will eventually been utilized in anaerobic oxidation to convert the aldehyde to either lactic acid or alchol. In higher vertebrate, under low oxygen pressure (such as during exercise in muscle) anerobic oxidation produces large amount of lactic acid but once oxygen is available lactic acid produced in muscle is sent to liver to regenerate glucose which will be send back to muscle for oxidative phosphorylation. This cyclic event is known as Cori cycle.

Growth and multiplication of host organisms used as expression system, requires a suitable biochemical and biophysical conditions. The biochemical (nutritional) conditions can be provided by the use of various nutrient media. Depending upon the special needs, different types of media have been developed for expression system to achieve growth, multiplication and over-expression of protein.

Growth Media for Bacterial Expression System

Bacterial expression systems are mainly utilized for protein over-expression because of its rapid growth rate, low cost, ease of high-cell-density fermentation, and availability of excellent genetic tools. Growth of bacterial expression system requires different types of media based on the requirement which can be divided into either complex or defined media. The complex media comprises of natural substances and rich in nutrients therefore suitable for culturing fastidious organism. On the other hand, defined media are simple and made up of known components put together in the required amounts.

Constituents	Source
Amino-Nitrogen	Peptone, protein hydrolysate, infusions and extracts
Growth Factors	Blood, serum, yeast extract or vitamins, NAD
EnergySources	Sugar, alcohols and carbohydrates
BufferSalts	Phosphates, acetates and citrates
Mineral salts and Metals	Phosphate, sulfate, magnesium, calcium, iron
Selective Agents	Chemicals, antimicrobials and dyes
Indicator Dyes	Phenol red, neutral red
Gelling agents	Agar, gelatin, alginate, silicagel

Common media constituents for Bacterial growth

Preparation of Bacterial Expression media: The composition of the selected bacterial expression media is given in the table. For preparation of bacterial media dissolve the components in 1 liter of distilled water. Cover the top of the flask with cotton plug or aluminium foil and autoclave the solution at 121° C for 20 minutes. The various antibiotics or nutrient supplement should be added to the media when the temperature is less than 50° C after autoclaving.

For making of solid media agar plates, 1.5% agar is added to the media and autoclaved. After autoclaving allow the media to cool up-to 50° C. Transfer the warm media into a petri dishes until it is 1/3 full, loosely close the lid and allow the media to solidify.

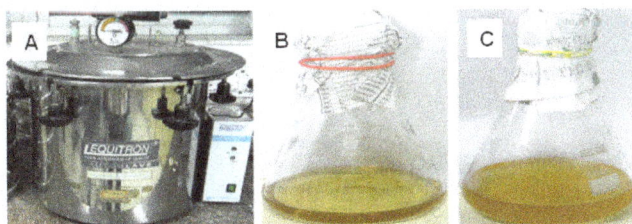

Equipments and media required for sterilization and growth of bacterial expression system. (A) autoclave (B)autoclaved LB broth (C) E. coli grown in LB broth.

Growth media	Compositions	Applications
M9 minimal media	0.6% disodium hydrogen phosphate 0.3% potassium dihydrogen phosphate, 0.05%, Sodium chloride 0.1% ammonium chloride	For cultivation and maintenance of *Escherichia coli* (*E. coli*) strains.
M63 minimal media	0.2% ammonium sulfate 1.36% potassium dihydrogen phosphate monobasic 0.00005% ferrous sulfate.7H$_2$O	For cultivation and maintenance of *E. coli* strains.
LB (Luria Bertani) Miller broth	1% peptone 0.5% yeast extract 1% NaCl	For *E. coli* growth; plasmid DNA isolation and protein production
LB (Luria Bertani) Lennox Broth	1% peptone 0.5% yeast extract 0.5% NaCl	For *E. coli* growth; plasmid DNA isolation and protein production
SOB medium	2% peptone 0.5% Yeast extract 10mM NaCl 2.5mM KCl, 20mM MgCl$_2$	To make high efficiency competent cells.
SOC medium	SOB + 20mM glucose	growth of competent cells.
2x YT broth (2x Yeast extract and Tryptone)	1.6% peptone 1% yeast extract 0.5% NaCl	Phage DNA production
Terrific Broth) medium	1.2% peptone, 2.4% yeast extract 72 mM K$_2$HPO$_4$ 17 mM KH$_2$PO$_4$ 0.4% glycerol	For protein expression and plasmid production.
Super Broth) medium	3.2% peptone, 2% yeast extract 0.5% NaCl	High yield plasmid DNA and protein production
TYGPN media	2% Tryptone, 1% Yeast extract, 1ml 80% Glycerol, 1%Potassium Nitrate, 0.5% Sodium Phosphate dibasic	For rapid growth of *E. coli*.

Composition of selected media for bacterial growth

Growth Media for Yeast Expression System

Yeast expression system offers advantages of speedy growth, easy genetic manipulation, low cost media with the characteristics of higher eukaryotic systems such as post translational modifications and secretory expression. Yeast expression system mainly utilizes *Saccharomyces cerevisiae* (*S. cerevisiae*) and *Pichia pastoris* (*P. pastoris*) strains for cloning and protein overexpression. For the growth, propagation and protein overexpression of these strains specific formulations and ingredients are required. The common type of yeast expression media are given in the table below.

Growth media	Composition (For 1 litre)	Applications
CSM Media	CSM (without tryptophan)0.74gm, Yeast Nitrogen Base6.66gm, AmmoniumSulfate5gm,Glucose20gm, Agar 20 gm	For making agar plates that enable the growth of Saccharomyces cerevisae MaV203 competent cells.
YPD Broth	10gm Yeast extract 20gm Bacto peptone 20gm Dextrose (glucose)	Commonly used yeast media for maintenance and propagation of *P. pastoris* and *S. cerevisae*.
YPGal	10gm Bacto Yeast Extract 20gm Bacto Peptone 100ml of 20% Galactose 15gm Bacto Agar	Standard medium for *S.cerevisae* omitting glucose repression
Standard Minimal Medium (SD) /Yeast Nitrogen Base (YNB)	6.7gm yeast nitrogen base with ammonium sulfate and without amino acids and 20gm dextrose plus any amino acids or nucleotides required for growth at ~50 ug/ml each).	Base medium for preparation of minimal and synthetic defined yeast media

Selected growth media for yeast expression system

Method of Preparation. Media preparation of yeast expression system is similar to the microbiology media. As per the media composition, constituents are in 950 ml of water and autoclave. Allow medium to cool to 50°C and then add 50ml of filtered sterile 40% dextrose (glucose) so that the final concentration become 2% . Adjust the final volume to 1 litre, if necessary.

Growth Media for Insect Cell Culture

In modern biotechnology insect cell culture is gaining a considerable attention for production of recombinant proteins. Insect cell systems provide improved target protein solubility and important post-translational modifications for increased activity. Baculovirus Expression Vector System (BEVS) is the well known system to utilize the insect cell lines for the production of recombinant proteins. The media used for insect cell culture is a complex mixture of Amino acid, Monosaccharide, Vitamin, Inorganic ion, trace elements, fetal bovine serum (FBS) and broad spectrum antibiotics. The popular culture media required for the growth of various insect lines are given in the table below.

Growth media	Compositions[#]	Applications
Grace's Insect medium supplemented	Unsupplemented media actalbumin hydrolysate yeastolate	Growth of *Spodeptera frugiperda* cells, Sf9 and Sf21 cell lines
Hink's TNM-FH Insect Medium	supplemented Grace's, 4.1 mM L-glutamine, 3.33g/Llactalbuminhydrolysate(LAH)	For the culture of cabbage looper, *Trichoplusia ni* cells
IPL-41 Insect Medium Modified	IPL-41 media Calcium chloride 200mM L-glutamine Sodium bicarbonate	Growth of *Spodeptera frugiperda* cells, Sf9 and Sf21 cell lines
TC-100 Medium	TC-100 Medium 200mM L-glutamine Sodium bicarbonate	For the production of baculovirus in lepidopteran cell lines.
Mitsuhashi/Maramorosh Insect Medium	Mitsuhashi/MaramoroshInsect Medium Sodium bicarbonate	For Mosquito cell culture especially *Aedes aegypticus*
Schneider's Drosophila Medium	Schneider's Drosophila Medium Calcium chloride 200mM L-glutamine Sodium bicarbonate	For the in vitro culture of Drosophila melanogaster cells and tissues

Selected growth media for insect cell culture

General method of preparation: Dissolve dry powder of medium in cell culture grade water (80% of the final volume). Mix powder until dissolved completely. Add required amount of other component, mix completely and adjust the pH to 6.9-7.3 using 1N NaOH or 1N HCl. Make up the final volume with cell culture grade water. Sterilize the solution using a 0.22μm membrane filter. Finally, aseptically add antibiotics and serum to the sterilized incomplete media.

Growth Media for Mammalian Cell Culture:

The media used for animal cell culture is a complex mixture of Amino acid, Monosaccharide, Vitamin, Inorganic ion, trace elements and broad spectrum antibiotics. The

other key ingredients of cell culture is a natural medium which may be ani mal body fluids or medium of tissue extraction, including plasma, serum, lymph, chicken embryos
leaching solution, etc. Serum, usually bovine or calf is the most commonly used natural
medium. Serum provide a similar osmotic pressure and pH as of body environment.
Serum enhances the cell attachment and provides extra nutrients, various hormones
like growth factor that promotes healthy growth of the cell. In order to monitor the
status of media, phenol red is added as a pH indicator. This will turn yellow if media
becomes acidic otherwise media at pH 7.2-7.4 remains red.

Components	Composition
DMEM	13.4 gm/ltr
Sodium bicarbonate	3.7gm/ltr
Fetal bovine serum (FBS)	10%
100X Antibiotic (Pencillin –Streptomycin)	1%

Recipe of mammalian cell culture complete media.

Prepartion of Cell Culture Medium

To explain the method of media preparation, we are taking the example of DMEM media. Measure 80 - 90% of the final volume of cell culture grade water. Add 13.4 gm dry
powder medium to the water and mix to dissolve it completely. For each liter of DMEM,
add 3.7g/L of sodium bicarbonate, mix completely and adjust the pH to 6.9 -7.1 using
1N NaOH or 1N HCl. Finally add cell culture grade water to the media to bring it to the
final volume. Sterilize the solution using a sterilized membrane filter with a pore size
of 0.22μm. Supplements, such as antibiotics and serum can be added to the sterilized
solution using aseptic technique.

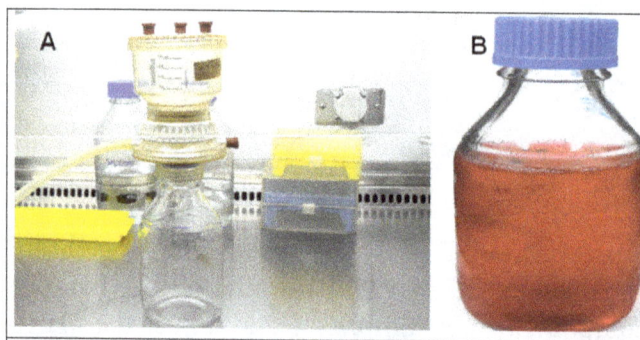

Sterilization of cell culture medium (a) filteration unit set up for
the filteration of DMEM media through 0.22μm filter. (b) complete media containing
10% serum and 1% pencillin–streptomycin antibiotic.

References

- Cuendet, M.A., Weinstein, H., and LeVine, M.V. (2016) "The allostery landscape: quantifying
  thermodynamic coupling in biomolecular systems". Journal of Chemical Theory and Computa-
  tion

- Lowenstein JM (1969). Methods in Enzymology, Volume 13: Citric Acid Cycle. Boston: Academic Press. ISBN 0-12-181870-5

- Zhang S, Bryant DA (December 2011). "The tricarboxylic acid cycle in cyanobacteria". Science. 334 (6062): 1551–3. PMID 22174252. doi:10.1126/science.1210858

- Krebs HA, Weitzman PD (1987). Krebs' citric acid cycle: half a century and still turning. London: Biochemical Society. p. 25. ISBN 0-904498-22-0

- Krebs, HA; Johnson, WA (April 1937). "Metabolism of ketonic acids in animal tissues.". The Biochemical journal. 31 (4): 645–60. PMC 1266984. PMID 16746382

- Ferre, P.; F. Foufelle (2007). "SREBP-1c Transcription Factor and Lipid Homeostasis: Clinical Perspective". Hormone Research. 68 (2): 72–82. PMID 17344645. doi:10.1159/000100426. Retrieved 2010-08-30. this process is outlined graphically in page 73

- Hilser, V. J.; Wrabl, J. O.; Motlagh, H. N. (2012). "Structural and energetic basis of allostery". Ann. Rev. Biophys. 41: 585–609. doi:10.1146/annurev-biophys-050511-102319

- Jones RC, Buchanan BB, Gruissem W (2000). Biochemistry & molecular biology of plants (1st ed.). Rockville, Md: American Society of Plant Physiologists. ISBN 0-943088-39-9

- Motlagh, Hesam N.; Wrabl, James O.; Li, Jing; Hilser, Vincent J. (2014). "The ensemble nature of allostery". Nature. 508: 331–340. doi:10.1038/nature13001

- Stryer L, Berg JM, Tymoczko JL (2002). "Section 18.6: The Regulation of Cellular Respiration Is Governed Primarily by the Need for ATP". Biochemistry. San Francisco: W.H. Freeman. ISBN 0-7167-4684-0

- Barnes SJ, Weitzman PD (June 1986). "Organization of citric acid cycle enzymes into a multienzyme cluster". FEBS Lett. 201 (2): 267–70. PMID 3086126. doi:10.1016/0014-5793(86)80621-4

3

Gene Sequencing: An Integrated Study

Gene sequence is arranged randomly and it becomes difficult to separate when they are also unknown. To resolve the problem, two ways have been developed to represent the genomic sequence, i.e. genomic library and cDNA library. Topics such as recombinant DNA and cloning among others have also been discussed. This chapter strategically encompasses and incorporates the major components and key concepts of biotechnology, providing a complete understanding.

Gene Sequence

Gene sequence are arranged in genome in a randon fashion and selecting or isolating a gene is a big task especially when the genomic sequences are not known. A small portion of genome is transcribed to give mRNA where as a major portion remained untranscribed. Hence, there are two ways to represent a genomic sequence information into the multiple small fragments in the form of a library: (1) Genomic library (2) cDNA library.

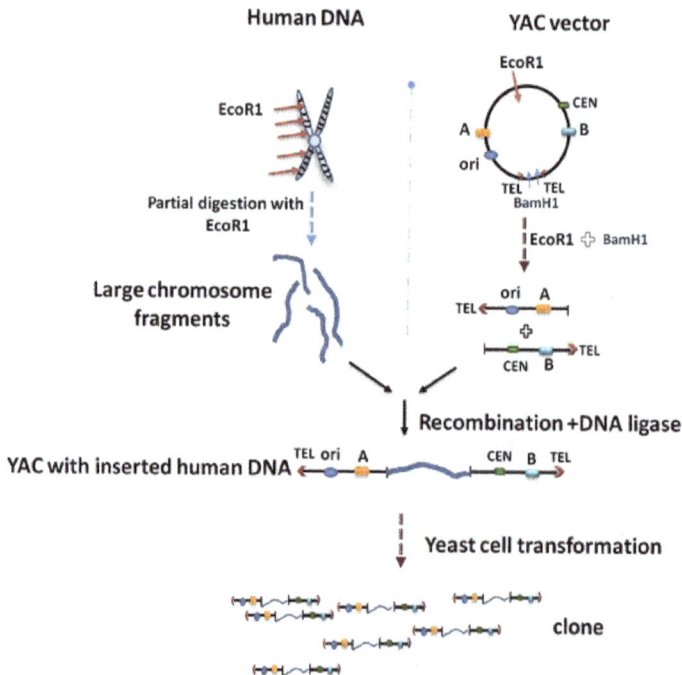

Contruction of Genomic library.

Preparation of Genomic Library-A genomic library represents complete genome in multiple clones containing small DNA fragments. Depending upon organism and size of genome, this library is either prepared in a bacterial vector or in yeast artificial chromosome (YAC). An outline of the construction of genomic library is given in the above figure. It has following steps:

- Isolation of genomic DNA

- Generation of suitable size DNA fragments

- Cloning in suitable vector system (depending on size)

- Transformation in suitable host .

1. Isolation of genomic DNA- Isolation of genomic DNA has following steps:

- Lysis of cells with detergent containing lysis buffer.

- Incubation of cells with digestion buffer containing protease-K, SDS to release genomic DNA from DNA-protein complex.

- Isolation of genomic DNA by absolute alchol precipitation.

- Purification of genomic DNA with phenol:chloroform mixture. Chloroform:phenol mixture has two phases, aquous phase and organic phase. In this step, phenol denatures the remaining proteins and keep the protein in the organic phase.

- Genomic DNA present in aqueous phase is again precipitated with absolute alchol.

- Genomic DNA is analyzed on 0.8% agarose gel and a good prepration of genomic DNA give an intact band with no visible smear.

2. Generation of suitable size fragments- Next step generation of genomic DNA into suitable small size fragments.

Restriction digestion: Genomic DNA can be digested with a frequent DNA cutting enzyme such as EcoR-I, BamH-I or sau3a to generate the random sizes of DNA fragments. The criteria to choose the restriction enzyme or pair of enzymes in such a way so that a reasonable size DNA fragment will be generated. As fragments are randomly generated and are relatively big enough, it is likely that each and every genomic sequence is presented in the pool. As size of the DNA fragment is large, complete genome will be presented in very few number of clones. In addition, genomic DNA can be fragmented using a mechanical shearing.

If a organism has a genome size of 2×10^7 kb and an average size of the fragment is 20kb, then no. of fragment, $n = 10^6$. In reality, this is the minimum number to represent a given fragment in the library where as the actual number is much larger. The probability (P) of finding a particular genomic sequence in a random library of N independent clone is as follows:

$$N=\ln (1-P)/\ln (1-1/n)$$

Where, N=number of clones, P=probability, n= size of average fragment size

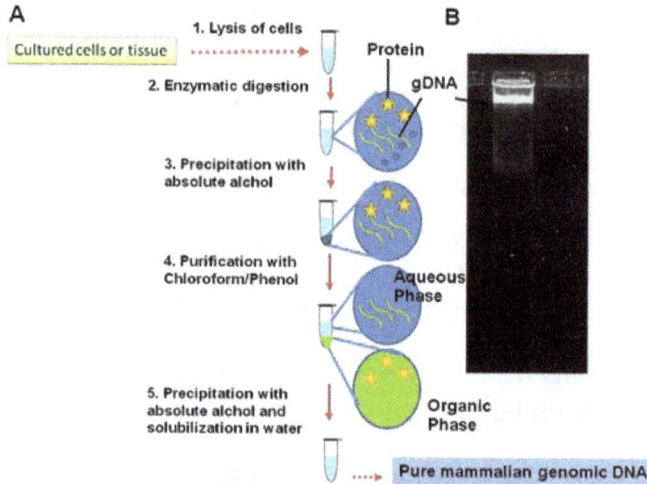

Genomic DNA isolation. (A) Different steps in genomic DNA isolation. (B) Agarose gel analysis of isolated genomic DNA.

3. Cloning into the suitable vector- The suitable vector to prepare the genomic library can be selected based on size of the fragment of genomic DNA and carrying capacity of the vector. Size of average fragment can be calculated from the Eq 10.1 and accordingly a suitable vector can be choosen. In the case of fragment generated by restriction enzyme, vector can be digested with the same enzyme and put for ligation to get clone. In the case of mechanical shearing mediated fragment generation, putting these fragment needs additional effort. In one of the approachs, a adopter molecule can be used to generate sticky ends, alternatively a endonuclease can be used to generate sticky ends.

4. Transformation to get colonies- Post ligation, clones are transformed in a suitable host to get colonies. A suitable host can be a bacterial strain or yeast.

S.NO	Vector	Insert Size (MB)
1	Plasmids	15
2	Phage lambda	25
3	Cosmids	45
4	Bacteriophage	70-100
5	Bacterial artificial chromosome (BAC)	120-300
6	Yeast artificial chromosome (YAC)	250-2000

Carrying capacity of different vectors

cDNA Library

A cDNA library is a combination of cloned cDNA (complementary DNA) fragments inserted into a collection of host cells, which together constitute some portion of the transcriptome of the organism and stored as a "library". cDNA is produced from fully transcribed mRNA found in the nucleus and therefore contains only the expressed genes of an organism. Similarly, tissue-specific cDNA libraries can be produced. In eukaryotic cells the mature mRNA is already spliced, hence the cDNA produced lacks introns and can be readily expressed in a bacterial cell. While information in cDNA libraries is a powerful and useful tool since gene products are easily identified, the libraries lack information about enhancers, introns, and other regulatory elements found in a genomic DNA library.

cDNA Library Construction

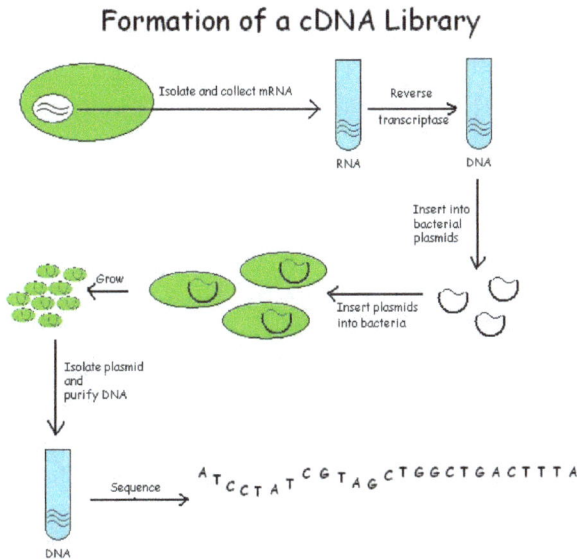

Formation of a cDNA library.

cDNA is created from a mature mRNA from a eukaryotic cell with the use of reverse transcriptase. In eukaryotes, a poly-(A) tail (consisting of a long sequence of adenine nucleotides) distinguishes mRNA from tRNA and rRNA and can therefore be used as a primer site for reverse transcription. This has the problem that not all transcripts, such as those for the histone, encode a poly-A tail.

mRNA Extraction

Firstly, the mRNA is obtained and purified from the rest of the RNAs. Several methods exist for purifying RNA such as trizol extraction and column purification. Column purification is done by using oligomeric dT nucleotide coated resins where only the mRNA having the poly-A tail will bind. The rest of the RNAs are eluted out. The mRNA is eluted by using eluting buffer and some heat to separate the mRNA strands from oligo-dT.

cDNA Construction

Once mRNA is purified, *oligo-dT* (a short sequence of deoxy-thymidine nucleotides) is tagged as a complementary primer which binds to the poly-A tail providing a free 3'-OH end that can be extended by reverse transcriptase to create the complementary DNA strand. Now, the mRNA is removed by using a RNAse enzyme leaving a single stranded cDNA (sscDNA). This sscDNA is converted into a double stranded DNA with the help of DNA polymerase. However, for DNA polymerase to synthesize a complementary strand a free 3'-OH end is needed. This is provided by the sscDNA itself by generating a *hairpin loop* at the 3' end by coiling on itself. The polymerase extends the 3'-OH end and later the loop at 3' end is opened by the scissoring action of S_1 *nuclease*. Restriction endonucleases and DNA ligase are then used to clone the sequences into bacterial plasmids.

The cloned bacteria are then selected, commonly through the use of antibiotic selection. Once selected, stocks of the bacteria are created which can later be grown and sequenced to compile the cDNA library.

cDNA Library Uses

cDNA libraries are commonly used when reproducing eukaryotic genomes, as the amount of information is reduced to remove the large numbers of non-coding regions from the library. cDNA libraries are used to express eukaryotic genes in prokaryotes. Prokaryotes do not have introns in their DNA and therefore do not possess any enzymes that can cut it out during transcription process. cDNA does not have introns and therefore can be expressed in prokaryotic cells. cDNA libraries are most useful in reverse genetics where the additional genomic information is of less use. Also, it is useful for subsequently isolating the gene that codes for that mRNA.

cDNA Library vs. Genomic DNA Library

cDNA library lacks the non-coding and regulatory elements found in genomic DNA. Genomic DNA libraries provide more detailed information about the organism, but are more resource-intensive to generate and maintain.

Cloning of cDNA

cDNA molecules can be cloned by using restriction site linkers. Linkers are short, double stranded pieces of DNA (oligodeoxyribonucleotide) about 8 to 12 nucleotide pairs long that include a restriction endonuclease cleavage site e.g. BamHI. Both the cDNA and the linker have blunt ends which can be ligated together using a high concentration of T4 DNA ligase. Then sticky ends are produced in the cDNA molecule by cleaving the cDNA ends (which now have linkers with an incorporated site) with the appropriate endonuclease. A cloning vector (plasmid) is then also cleaved with the appropriate endonuclease. Following "sticky end" ligation of the insert into the vector the resulting recombinant DNA molecule is transferred into *E. coli* host cell for cloning.

Identification and isolation of a particular gene is essential for development of the biotechnology applications. With the availability of genomic sequences the task is getting easier day-by-day. The approaches we would like to discuss today is to identify and isolate a gene fragment from an organism. We have discussed presenting genomic DNA either not associated with the production of protein (Genomic Library) or responsible for production of protein (cDNA library). We will discuss screening and isolation of gene with known structural (DNA sequence) or functional attributes (enzyme activity or particular antigenic epitope).

There are 3 different searchable criteria to identify a particular gene from an organism:

1. DNA sequence-This properties can be use to search both genomic library and cDNA library to identify the gene.

2. Expression of a particular protein with immunogenic epitope-This property can be partially useful to screen genomic library due to truncation of a full gene or no expression of a gene fragment. But this approach suits well to screen cDNA clones.

3. Enzymatic activity- This property exploits the ability of a protein fragment to exhibits enzymatic activity. It is useful for the screening of cDNA library but not much for genomic library.

Screening by DNA Hybridization

DNA sequence information can be exploited with a general rule that nucleotide present in a DNA sequence provides a specificity due to unique base pairing preference of nucleotides. "A" is always making base pairing with "T" and "G" is making base pairing with "C".

$$A = T$$

$$G \equiv C$$

As a result a particular DNA sequence can be identified by a complementary single stranded DNA sequence. The DNA sequence used for this purpose is called as "Probe". After-wards the position of probe can be identified by a suitable detection system. The position of probe is the actual site of desirable clone of containing specific sequence. This complete procedure of colony hybridization is given in the figure below and it has following steps:

1. Preparation of suitable radioactive probe.

2. Preparation of replica plate

3. Transfer of colonies on nitrocellulose membrane.

4. Hybridization with a specific probe.

5. Washing and development of membrane by autoradiography.

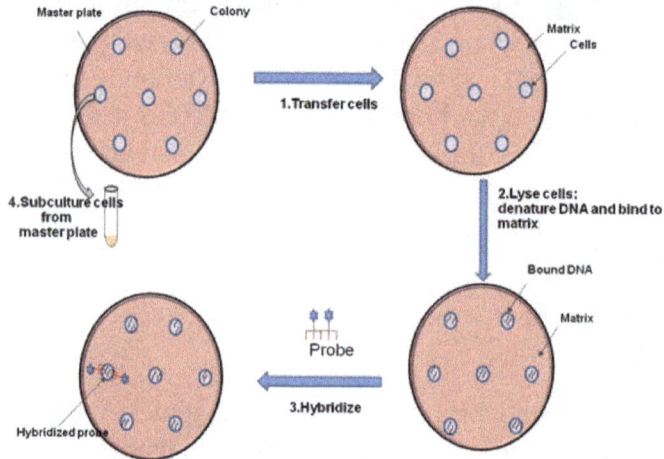

Screening a library with a radioactive probe by colony hybridization.

1. Preparation of radioactive probe. There are two different method used to label a single stranded DNA probe either at terminal or through the sequence.

A. Random primer method- In this method, a random primer is used to anneal to the template and then a PCR reaction is performed in the presence of radio-labeled nucleotide. After PCR, newly synthesized DNA strand is labeled with radioactive nucleotide. The whole process is given in the figure, and it has following steps-

    •   The source double stranded DNA is denatured to generate the single stranded DNA template.

    •   A random primer is added and allowed it to anneal to the template strand. It will anneal to the random position through out the sequence at multiple places.

    •   Primer will anneal to the template strand and now klenow will start the synthesis of new DNA strand.

    •   Newly synthesized DNA will give short stretches of multiple labeled DNA probes.

B. Terminal transferease- In this method, a terminal transferase enzyme will label the probe at the ends to the last nucleotide of the probe. Probe is incubated with the labeled nucleotide and terminal transferase enzyme will add the labeled nucleotide at the end. A partial purification with gel filtration column will give labeled primer.

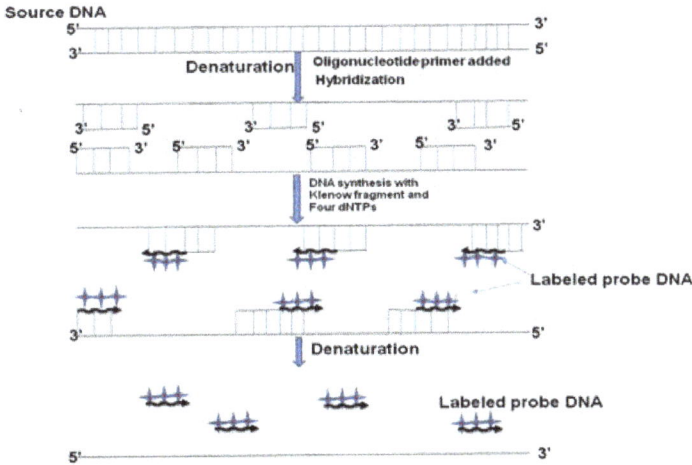

Preparation of radioactive probe by random primer method.

2. Preparation of Replica plate- As original genomic or cDNA library is precious and will be consumed in later stage, all procedure is performed with the replica plate containing clones in an identical manner.

3. Blotting-The clone is transferred on a nitrocellulose membrane with retaining identical pattern of colonies on master plate. The cells on the membrane are lysed and released DNA is denatured, deproteinated and allowed to bind the membrane.

4. Hydridization-A labeled probe prepared in step 1 will be added. Probe will binds to the target DNA due to base pairing. The membrane is washed to remove unbound probe.

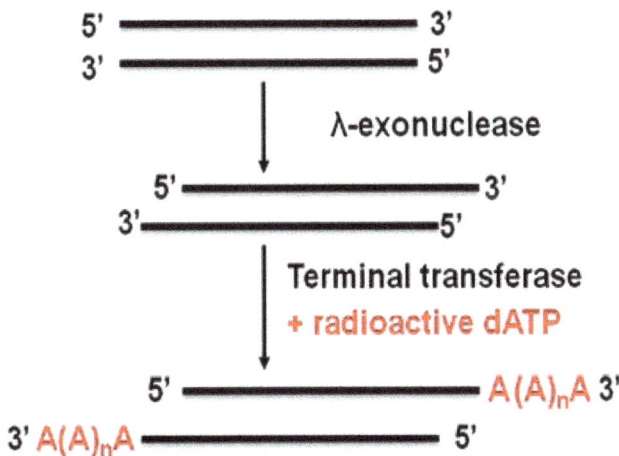

Preparation of radioactive probe by terminal transferase method.

5. Development of blot (Autoradiography)-The position of labeled probe is detected by autoradiogram.The position of signal on membrane can be matched with the master plate to get location of corresponding colony.

Screening by Immunological Methods

This method is based on the specificity of antibody towards its antigenic epitope present on the protein expressed in a particular clone. A number of disease associated gene have been identified by this method. Due to increased expression or unique expression of a particular protein in a disease condition, patient body develops antibody against it. The developed antibody is available to use to identify the protein expressing clone. This method has following steps:

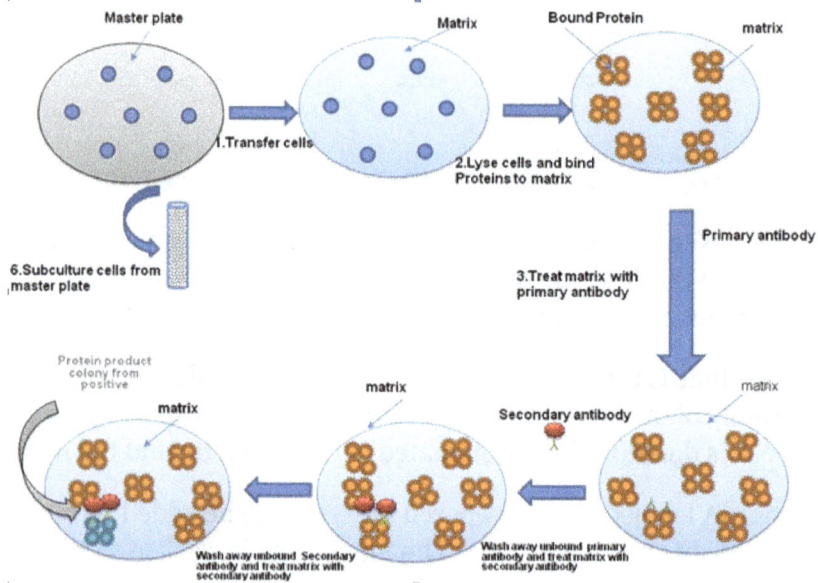

Sceening by immunological methods

1. Preparation of Replica plate- As original genomic or cDNA library is precious and will be consumed in later stage, all procedure is performed with the replica plate containing clones in a indentical manner.

2. Blotting-The clone is transferred on a nitrocellulose membrane to get similar pattern of colonies on master plate. The cells on the membrane are lysed and released protein is denatured, and allowed to bind the membrane.

3. Treatment with primary antibody-The membrane is incubated with antibody having immunoreactivity towards a particular protein. The primary antibody will binds to the target protein due to exclusive specificity towards antigen. The membrane is washed to remove unbound primary antibody.

4. Treatment with secondary antibody-The membrane is incubated with secondary antibody recognizing primary antibody. Secondary antibody is labeled with an enzyme (Horse raddish peroxidase or alkaline phosphatase) to use to give readable signal. The secondary antibody will binds to the primary antibody and will allow

to detect the location of primary antibody. The membrane is washed to remove un-bound secondary antibody.

Why enzyme labeled secondary antibody is used instead of labeled primary antibody ?

Development of blot-The position of secondary antibody is detected by performing enzymatic activity.The position of signal on membrane can be matched with the master plate to get location of corresponding colony.

Screening by Enzymatic Activity

This method is based on the ability of protein to exhibit an enzymatic activity. This method is not very specific but allow us to identify a class of protein with known enzymatic activity.

Isolation of gene-Once the position of a clone is known, it is extracted from the master plate and plasmid is isolated. In few cases, clone is further diluted to check the homogeneity of clone. The purity of the clone and presence of clone is further tested with a PCR using sequence specific primers.

Competent Cells

The delivery of DNA into the host is required for generation of genetically modified organism. DNA delivery to host is a 3 stage process, DNA sticking to the host cell, internalization and release into the host cell. As a result, it depends on 2 parameters-

Surface chemistry of host cell-Host cell surface charges either will attract or repell DNA as a result of opposite or similar charges. Presence of cell wall (in the case of bacteria, fungus and plant) causes additional physical barrier to the up-take and entry of DNA.

Charges on DNA-Negative charge on DNA modulates interaction with the host cell especially cell surface.

Modulation of these two properties is achieved in different methods to deliver DNA into the host cell.

Lab Experiment: Preparation of Chem Cally (CaCl$_2$) Treated E. coli Competent Cells.

Background Information: Natural ability of a cell (either bacterium/yeast or mammalian cell) to take up cell free DNA present in extracellular environment is low and only 1% cells are capable to take DNA. Hence, a number of (chemical/physical) treatments make the bacteria competent to take DNA through transformation. A list of agent used for different organism is given in the table below.

Bacterial Strains	Competent agents
Streptococcus pneumoniae	mitomycin C, fluoroquinolone
In *B. subtilis*	UV light
Helicobacter pylori	ciprofloxacin
Legionella pneumophila	mitomycin C, norfloxacin, ofloxacin, nalidixic acid, bicyclomycin, hydroxyurea, UV light
E.Coli	Calcium chloride, Rubidium Chloride
The most popular reagent for making E.coli competent cell is calcium chloride.	

List of selected agent as potential to make cell competent

Material Required

1. DH5α Host cells stock.

2. Luria Bertani medium: The composition of the luria Bertani and other bacterial expression media is given in the table below. For preparation of media dissolve the components in 1 liter of distilled water. Cover the top of the flask with cotton plug or aluminium foil and autoclave the solution at 121°C for 20 minutes. The various antibiotics or nutrient supplement should be added to the media when the temperature is less than 50°C .

3. 0.1 M Sterile $MgCl_2$, 0.1M sterile $CaCl_2$

Growth media	Compositions	Applications
M9 minimal media	0.6% disodium hydrogen phosphate 0.3% potassium dihydrogen phosphate, 0.05%, Sodium chloride 0.1% ammonium chloride	For cultivation and maintenance of *Escherichia coli (E. coli)* strains.
M63 minimal media	0.2% ammonium sulfate 1.36% potassium dihydrogen phosphate monobasic 0.00005% ferrous sulfate.7H_2O	For cultivation and maintenance of *E. coli* strains.
LB (Luria Bertani) Miller broth	1% peptone 0.5% yeast extract 1% NaCl	For *E.coli* growth; plasmid DNA isolation and protein production
LB (Luria Bertani) Lennox Broth	1% peptone 0.5% yeast extract 0.5% NaCl	For *E.coli* growth; plasmid DNA isolation and protein production
SOB medium	2% peptone 0.5% Yeast extract 10mM NaCl 2.5mM KCl, 20mM $MgCl_2$	To make high efficiency competent cells.
SOC medium	SOB + 20mM glucose	growth of competent cells.
2xYT broth (2x Yeast extract and Tryptone)	1.6% peptone 1% yeast extract 0.5% NaCl	Phage DNA production
Terrific Broth) medium	1.2% peptone, 2.4% yeast extract 72 mM K_2HPO_4 17 mM KH_2PO_4 0.4% glycerol	For protein expression and plasmid production.
Super Broth) medium	3.2% peptone, 2% yeast extract 0.5% NaCl	High yield plasmid DNA and protein production
TYGPN media	2% Tryptone, 1% Yeast extract, 1ml 80% Glycerol, 1%Potassium Nitrate, 0.5% Sodium Phosphate dibasic	For rapid growth of *E. coli*.

Composition of Selected media for bacterial Growth

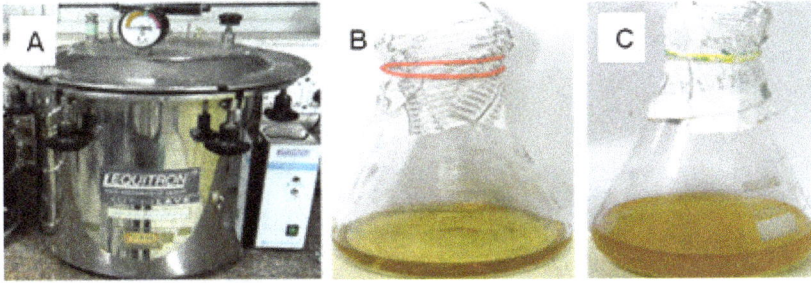

Equipments and media required for sterilization and growth of bacterial expression system. (A) autoclave (B) autoclaved LB broth (C) E. coli grown in LB broth.

Methods :

1. Bacterial Culture- The growth stage of the bacteria has a significant impact for its ability to take up foreign DNA. The bacterium at log phase is more active and efficient to perform DNA damage and repair than stationery phase. As a result, it is preferred to use a bacteria of log phase for making competent cells for transformation.

2. Preparation of Competent Cell-Bacteria is incubated with divalent cation (Calcium chloride,Manganese chrloride or Rubidium chloride) for 30mins at 4°C. During this process, cell wall of treated bacteria is swell and it gather factors required for intake of DNA docked on the plasma membrane. The whole process of E. coli competent cells preparation is as follows:

1. Inoculate single colony into the 100ml LB media and allow the cells to grow at 37°C, 180rpm until OD600 nm reaches to the 0.4-0.6.

2. Centrifuge the bacterial culture at 4000 rpm, at 4°C 10 min. Discard the supernatant.

3. Resuspend the cell pellet gently first in 1-2 ml and then in 10 ml of ice-cold 0.1 M $MgCl_2$.

4. Centrifuge the bacterial suspension at 4000 rpm, at 4°C 10 min. Discard the supernatant.

5. Resuspend cells gently in 3.0 ml of ice-cold 0.1 M $CaCl_2$.

6. Incubate the cell suspension on ice for additional 2hrs.

7. Centrifuge the bacterial suspension at 4000 rpm, at 4°C 10 min. Discard the supernatant.

8. Resuspend cells gently in 3.0 ml of ice-cold 0.1 M $CaCl_2$ containing 10% glycerol and store in small aliquot (100µl) at -80oC. The cells can be used for transformation.

Transformation

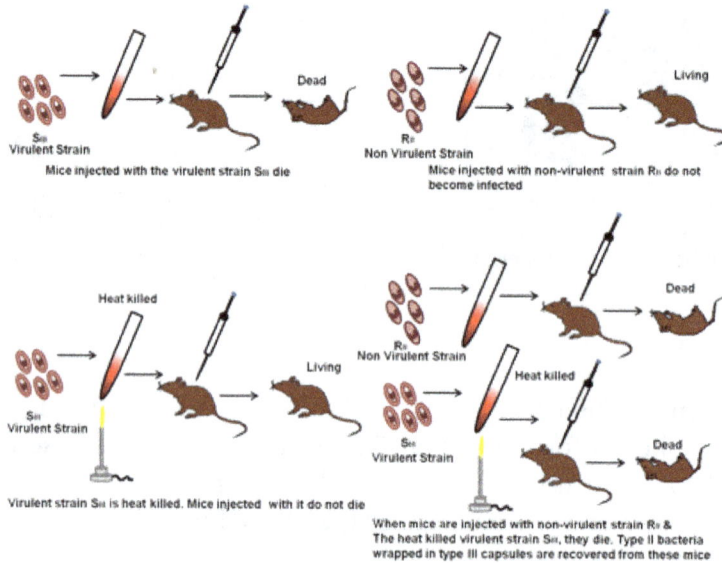

Discovery of Transformation

Transformation- it is the natural process, through which bacterial population transfer the genetic material to acquire phenotypic features. The event of transformation was first time demonstrated by Frederick Griffith in 1928. The schematic presentation of the experiment is given in the figure above. Griffith has used two different *Streptococcus pneumonia* strains, virulent (S, causes disease and death of mice) and avirulent (R, incapable of causing disease or death of mice). In a simple experiment he injected 4 different combination of bacterial mixture, (1) live S, (2) heat killed S, (3) live R, (4) mixture of live R and heat killed S in to the mice. The observation indicates that live S has killed the mice where as mice were healthy with heat killed S or live R. Surprisingly, mice injected with mixture of live R with heat killed S were found dead, and bacteria isolated from these dead mice were virulent. Based on these observations, Griffith hypothesized the existence of a transforming agent (Protein, DNA) being transferred from heat killed virulent strain to the avirulent strain and proposed the concept of transformation. Later, Oswald has proved that the transforming factor is DNA rather than protein.

Lab Experiment: Transformation PET23a into the E. coli and Calculate Transformation Efficiency.

Background Information: Transformation is the process by which cell free DNA is taken up by another bacteria. The principle steps of transformation are given inthe figure below. The DNA from donor bacteria binds to the competent recipient cell and DNA enters into the cell. The DNA enters into the recipient cell through a uncharacterized mechanism. The DNA is integrated into the chromosomal DNA through a homologous recombination. Naturally transformation is common between closely related species only.

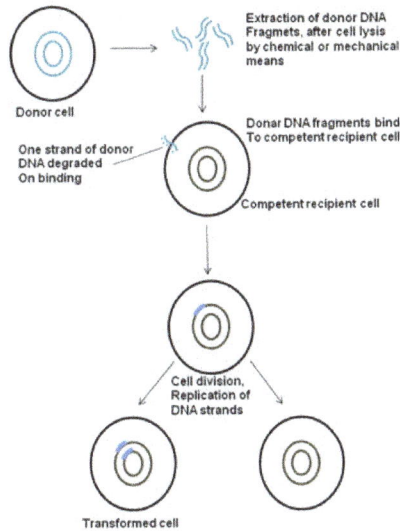

Principle steps in transformation.

Material and Equipments:

1. E. Coli Competent Cells

2. Water Bath

3. Luria Bertani Media

4. LB-Agar Plate

5. Incubator

Procedure : The outline of the procedure involved in calcium chloride mediated transformation is given in the figure below.

1. On the day of transformation, competent cells are incubated with DNA or circular plasmid containing appropriate resistance gene such as ampicillin resistance gene for 30mins on ice.

2. Heat Shock-Competent cells are given a brief heat shock (42° C for 90 sec) to relax the cell wall and high temperature stress causes upregulation of the factor responsible for DNA recombination and repair.

3. A chilled bacterial media is added for faster recovery of transformed cells.

4. it is plated on the solid media with appropriate antibiotics such as ampicillin and allowed to grow for another 18-24 hrs.

5. Transformed cells with appropriate resistance will grow and give colony.

Observation: The transformed LB agar plate in the figure has 1488 colonies.

Calculation of the transformation efficiency: If 10ng of plasmid gives ~1488 colonies. Then 1μg of plasmid transform 1488 x10^2=1.488x 10^5 colonies

Therefore the transformation efficiency was found to be -1.4 x 10^6 colonies.

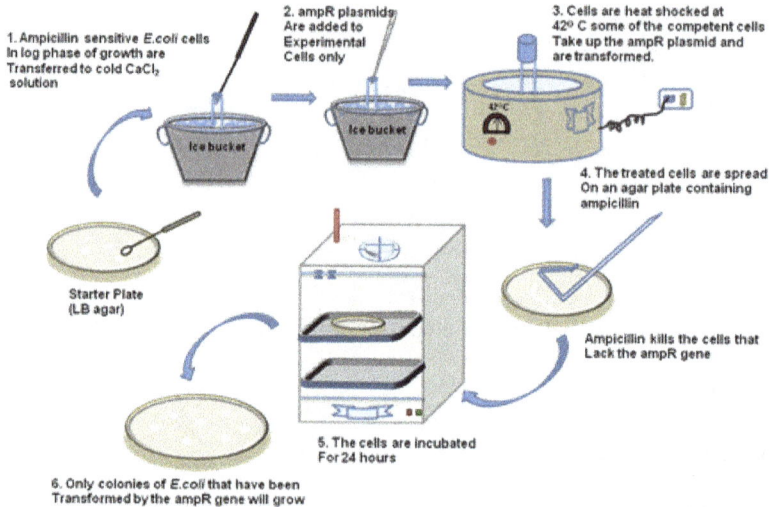

1. Ampicillin sensitive *E.coli* cells In log phase of growth are Transferred to cold $CaCl_2$ solution

2. ampR plasmids Are added to Experimental Cells only

3. Cells are heat shocked at 42º C some of the competent cells Take up the ampR plasmid and are transformed.

Ice bucket

Ice bucket

4. The treated cells are spread On an agar plate containing ampicillin

Starter Plate (LB agar)

Ampicillin kills the cells that Lack the ampR gene

5. The cells are incubated For 24 hours

6. Only colonies of *E.coli* that have been Transformed by the ampR gene will grow

Steps in bacterial transformation by $CaCl_2$ method.

Lab Experiment: Transforming the vector into the yeast.

Background Information: The surface chemistry of yeast cells are different from the E. coli and as a result different methods to deliver DNA.

1. Lithium Acetate/ssDNA/PEG Method: In this method, yeast cells are incubated with a transformation mixture of lithium acetate, PEG 3500, single stranded carrier DNA and foreign plasmid at 42°C for 40mins. The purpose of adding carrier DNA is to block the non-specific sites on cell wall and made plasmid available for uptake. Post-transformation, cells are pelleted to remove transformation mixture and re-suspended in 1ml water. It is plated on a solid media with an appropriate selection pressure such as antibiotics.

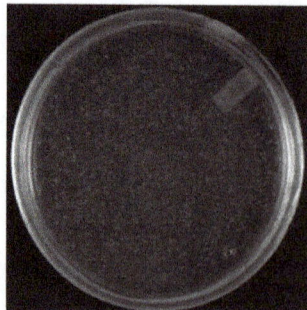

Transformation by $CaCl_2$ method.

2. Spheroplast Transformation Method: In this method, yeast cell wall is removed partially to produce spheroplast. Spheroplasts are very fragile for osmotic shock but are competent to takes up free DNA at high rate. In addition, polyethyl glycol (PEG) is used to facilitate deposition of plasmid and carrier DNA on cell wall for easier uptake. The mechanism of DNA uptake in yeast is not very clear. A schematic of spheroplast method is given in the figure below. In the spheroplast method, yeast cells are incubated with zymolyase to partially remove cell wall to produce spheroplast. (2) They are collected by centrifugation and incubated with carrier DNA and plasmid DNA for 10mins at room temperature. (3) It is now treated with PEG and calcium for 10mins with gentle shaking. (4) Transformed spheroplast are plated on selective solid media and incubated on 30°C for 4 days.

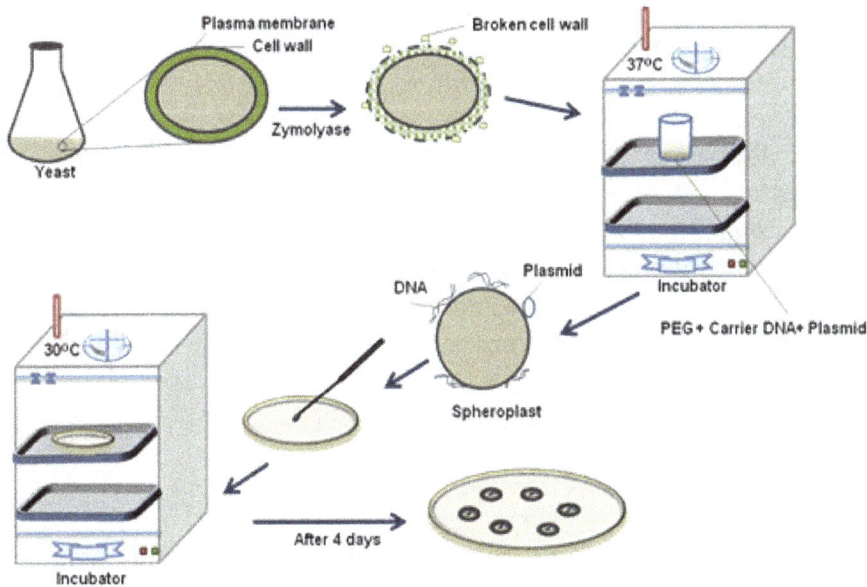

Steps in yeast transformation by sphereplast method.

Lab Experiment: Transfect GFP Into the COS-7 Cells using ion for Mammalian Cells.

Background Information: mammalian cell membrane surface chemistry, intracellular comparatmentization and uptake mechanism is different from the prokaryotic cells or yeast. Hence specialized methods have been developed to suit mammalian cells. There are 5 major strategies to deliver the DNA in mammalian cells:

1. Chemical transfection techniques-The principle behind the chemical transfection technique is to coat or complex the DNA with a polymeric compound to a reasonable size precipitate. It facilitates the interaction of the precipitate with the plasma membrane and uptake through endocytosis. There are multiple chemical compounds have been discovered which can be able to make complex and deliver DNA into the mammalian cell.

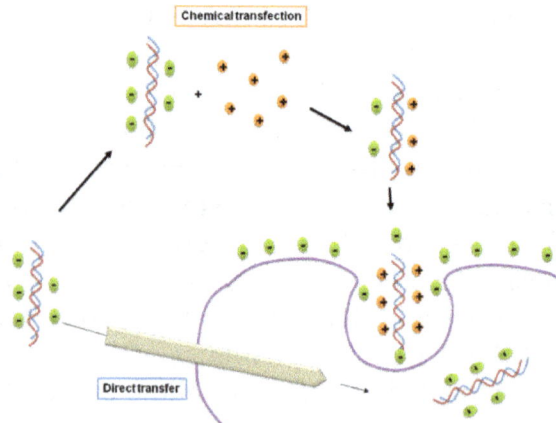

Mechanism of chemical method mediated DNA delivery in animal cells.

2. Calcium Phosphate method-In this method, DNA is mixed with calcium chloride in phosphate buffer and incubated for 20mins. Afterwards, transfection mixture is added to the plate in dropwise fashion. DNA-calcium phosphate complex forms a precipitate and deposit on the cells as a uniform layer. The particulate matter is taken up by endocytosis into the internal storage of the cell. The DNA is then escapes from the precipitate and reach to nucleus through a unknown mechanism. This method suited to the cell growing in monolayer or in suspension but not for cells growing in clumps. But the technique is inconsistent and the successful transfection depends on DNA-phosphate complex particle size and which is very difficult to control.

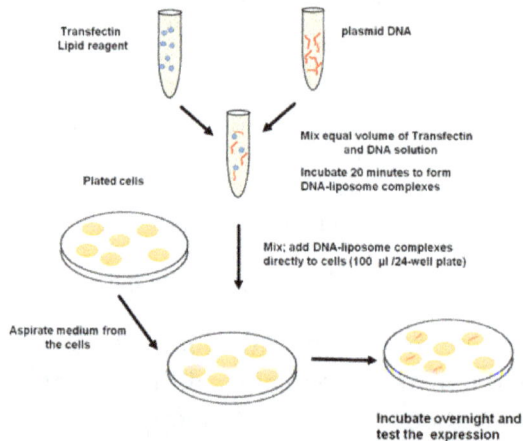

Transfection of animal cell with tranfectin (polyplexes)

3. Polyplexes method- The disadvantage of calcium phosphate method is severe physical damage to the cellular integrity due to particulate matter settling on the cell. It results in reduced cellular viability and cyto-toxicity to the cell. An alternate method was evolved where DNA was complexed with chemical agent to form soluble precipitate (polyplexes) through electrostatic interaction with DNA. A number of polycationic carbohydrate (DEAE-Dextran), positively charged cationic lipids

(transfectin), polyamines (polyethylenimines) etc are used for this purpose. The soluble aggregates of DNA with polycationic complex is readily been taken up by the cell and it reaches to the nucleus for expression.

4. Liposome and lipoplex method-Another approach of DNA transfection in animal cell is to pack the DNA in a lipid vesicle or liposome. In this approach, DNA containing vesicle will be fused with the cell membrane and deliver the DNA to the target cell. Preparation of liposome and encapsulating DNA was a crucial step to achieve good transfection efficiency. Liposome prepared with the cationic or neutral lipid facilitates DNA binding to form complex (lipoplex) and allow uptake of these complexes by endocytosis. The lipoplex method was applicable to a wide variety of cells, and found to transfect large size DNA as well. Another advantage of liposome/lipoplexes is that with the addition of ligand in the lipid bilayer, it can be used to target specific organ in the animal or a site within an organ.

5. Transduction (Virus mediated)- Viral particle has a natural tendency to attack and deliver the DNA into the eukaryotic cells. As discussed previously, cloning gene of interest in to the viral vectors is a innovative way to deliver the DNA into the host cell. If the recombination sequences are available, the delivered DNA is integrated into the host and replicate. Virus has essential components for expression of proteins required for DNA replication, RNA polymerase and other ligand for attachment onto the host cell. In addition, it has additional structural components to regulate infection cycle. The virus vector contains cassettes to perform all these functions then it is fully sufficient to propagate independently. Few virus strains may cause disease if their propagation will be uncontrolled. A mechanism has been devised to keep a check on the uncontrolled propagation of virus in cell. Few crucial structural blocks are placed on another helper plasmid, in this case virus propagate only if helper plasmid has been supplied along with the viral vector. This particular arrangement is made with the virus strains which can cause disease after integrating into the genome such as lentivirus.

Recombinant DNA

Recombinant DNA (rDNA) molecules are DNA molecules formed by laboratory methods of genetic recombination (such as molecular cloning) to bring together genetic material from multiple sources, creating sequences that would not otherwise be found in the genome. Recombinant DNA is possible because DNA molecules from all organisms share the same chemical structure. They differ only in the nucleotide sequence within that identical overall structure.

Recombinant DNA is the general name for a piece of DNA that has been created by the combination of at least two strands. Recombinant DNA molecules are sometimes called chimeric DNA, because they can be made of material from two different species, like the mythical chimera. R-DNA technology uses palindromic sequences and leads to the production of sticky and blunt ends.

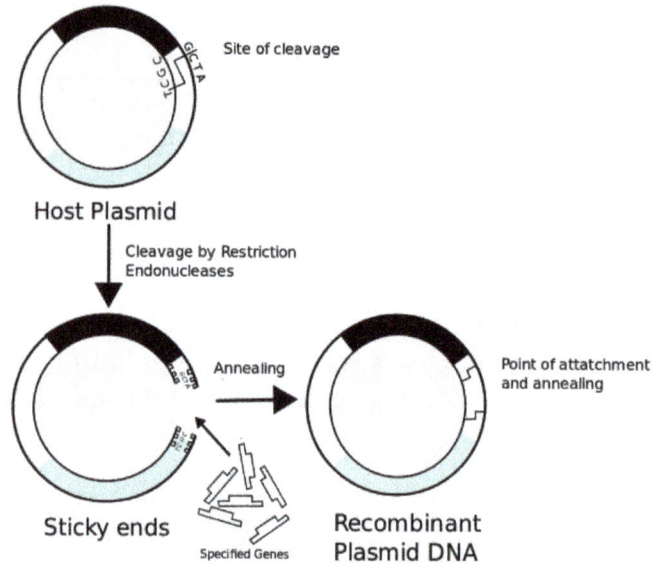

Construction of recombinant DNA, in which a foreign DNA fragment is inserted into a plasmid vector. In this example, the gene indicated by the white color is inactivated upon insertion of the foreign DNA fragment.

The DNA sequences used in the construction of recombinant DNA molecules can originate from any species. For example, plant DNA may be joined to bacterial DNA, or human DNA may be joined with fungal DNA. In addition, DNA sequences that do not occur anywhere in nature may be created by the chemical synthesis of DNA, and incorporated into recombinant molecules. Using recombinant DNA technology and synthetic DNA, literally any DNA sequence may be created and introduced into any of a very wide range of living organisms.

Proteins that can result from the expression of recombinant DNA within living cells are termed *recombinant proteins*. When recombinant DNA encoding a protein is introduced into a host organism, the recombinant protein is not necessarily produced. Expression of foreign proteins requires the use of specialized expression vectors and often necessitates significant restructuring by foreign coding sequences.

Recombinant DNA differs from genetic recombination in that the former results from artificial methods in the test tube, while the latter is a normal biological process that results in the remixing of existing DNA sequences in essentially all organisms.

Creation

Molecular cloning is the laboratory process used to create recombinant DNA. It is one of two widely used methods, along with polymerase chain reaction (PCR), used to direct the replication of any specific DNA sequence chosen by the experimentalist. There are two fundamental differences between the methods. One is that molecular cloning involves replication of the DNA within a living cell, while PCR replicates DNA in the test tube, free of living cells. The other difference is that cloning involves cutting and pasting DNA sequences, while PCR amplifies by copying an existing sequence.

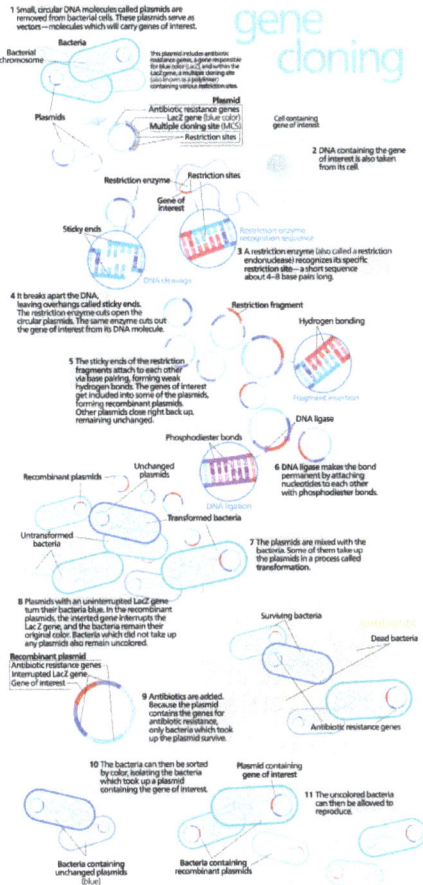

gene
cloning

1 Small, circular DNA molecules called plasmids are removed from bacterial cells. These plasmids serve as vectors—molecules which will carry genes of interest.

Bacteria

Bacterial chromosome

This plasmid includes antibiotic resistance genes, a gene responsible for blue color (LacZ) and within the LacZ gene, a multiple cloning site (also known as a polylinker) containing various restriction sites.

Plasmid
Antibiotic resistance genes
LacZ gene (blue color)
Multiple cloning site (MCS)
Restriction sites

Plasmids

Cell containing gene of interest

2 DNA containing the gene of interest is also taken from its cell.

Restriction enzyme   Restriction sites

Gene of interest

Sticky ends

Restriction enzyme recognition sequence

3 A restriction enzyme (also called a restriction endonuclease) recognizes its specific restriction site—a short sequence about 4–8 base pairs long.

DNA cleavage

4 It breaks apart the DNA, leaving overhangs called sticky ends. The restriction enzyme cuts open the circular plasmid. The same enzyme cuts out the gene of interest from its DNA molecule.

Restriction fragment

Hydrogen bonding

5 The sticky ends of the restriction fragments attach to each other via base pairing, forming weak hydrogen bonds. The genes of interest get included into some of the plasmids, forming recombinant plasmids. Other plasmids close right back up, remaining unchanged.

Fragment insertion

DNA ligase

Phosphodiester bonds

Recombinant plasmids   Unchanged plasmids

DNA ligation

6 DNA ligase makes the bond permanent by attaching nucleotides to each other with phosphodiester bonds.

Transformed bacteria

Untransformed bacteria

7 The plasmids are mixed with the bacteria. Some of them take up the plasmids in a process called transformation.

8 Plasmids with an uninterrupted LacZ gene turn their bacteria blue. In the recombinant plasmids, the inserted gene interrupts the LacZ gene, and the bacteria remain their original color. Bacteria which did not take up any plasmids also remain uncolored.

Surviving bacteria

Dead bacteria

Recombinant plasmid
Antibiotic resistance genes
Interrupted LacZ gene
Gene of interest

9 Antibiotics are added. Because the plasmid contains the genes for antibiotic resistance, only bacteria which took up the plasmid survive.

Antibiotic resistance genes

10 The bacteria can then be sorted by color, isolating the bacteria which took up a plasmid containing the gene of interest.

Plasmid containing gene of interest

11 The uncolored bacteria can then be allowed to reproduce.

Bacteria containing unchanged plasmids (blue)

Bacteria containing recombinant plasmids

Formation of recombinant DNA requires a cloning vector, a DNA molecule that replicates within a living cell. Vectors are generally derived from plasmids or viruses, and represent relatively small segments of DNA that contain necessary genetic signals for replication, as well as additional elements for convenience in inserting foreign DNA, identifying cells that contain recombinant DNA, and, where appropriate, expressing the foreign DNA. The choice of vector for molecular cloning depends on the choice of host organism, the size of the DNA to be cloned, and whether and how the foreign DNA is to be expressed. The DNA segments can be combined by using a variety of methods, such as restriction enzyme/ligase cloning or Gibson assembly.

In standard cloning protocols, the cloning of any DNA fragment essentially involves seven steps: (1) Choice of host organism and cloning vector, (2) Preparation of vector DNA, (3) Preparation of DNA to be cloned, (4) Creation of recombinant DNA, (5)

Introduction of recombinant DNA into the host organism, (6) Selection of organisms containing recombinant DNA, and (7) Screening for clones with desired DNA inserts and biological properties.

Expression

Following transplantation into the host organism, the foreign DNA contained within the recombinant DNA construct may or may not be expressed. That is, the DNA may simply be replicated without expression, or it may be transcribed and translated and a recombinant protein is produced. Generally speaking, expression of a foreign gene requires restructuring the gene to include sequences that are required for producing an mRNA molecule that can be used by the host's translational apparatus (e.g. promoter, translational initiation signal, and transcriptional terminator). Specific changes to the host organism may be made to improve expression of the ectopic gene. In addition, changes may be needed to the coding sequences as well, to optimize translation, make the protein soluble, direct the recombinant protein to the proper cellular or extracellular location, and stabilize the protein from degradation.

Properties of Organisms Containing Recombinant DNA

In most cases, organisms containing recombinant DNA have apparently normal phenotypes. That is, their appearance, behavior and metabolism are usually unchanged, and the only way to demonstrate the presence of recombinant sequences is to examine the DNA itself, typically using a polymerase chain reaction (PCR) test. Significant exceptions exist, and are discussed below.

If the rDNA sequences encode a gene that is expressed, then the presence of RNA and/or protein products of the recombinant gene can be detected, typically using RT-PCR or western hybridization methods. Gross phenotypic changes are not the norm, unless the recombinant gene has been chosen and modified so as to generate biological activity in the host organism. Additional phenotypes that are encountered include toxicity to the host organism induced by the recombinant gene product, especially if it is over-expressed or expressed within inappropriate cells or tissues.

In some cases, recombinant DNA can have deleterious effects even if it is not expressed. One mechanism by which this happens is insertional inactivation, in which the rDNA becomes inserted into a host cell's gene. In some cases, researchers use this phenomenon to "knock out" genes to determine their biological function and importance. Another mechanism by which rDNA insertion into chromosomal DNA can affect gene expression is by inappropriate activation of previously unexpressed host cell genes. This can happen, for example, when a recombinant DNA fragment containing an active promoter becomes located next to a previously silent host cell gene, or when a host cell gene that functions to restrain gene expression undergoes insertional inactivation by recombinant DNA.

Uses

A group of GloFish fluorescent fish

Recombinant DNA is widely used in biotechnology, medicine and research. Today, recombinant proteins and other products that result from the use of DNA technology are found in essentially every western pharmacy, doctor's or veterinarian's office, medical testing laboratory, and biological research laboratory. In addition, organisms that have been manipulated using recombinant DNA technology, as well as products derived from those organisms, have found their way into many farms, supermarkets, home medicine cabinets, and even pet shops, such as those that sell GloFish and other genetically modified animals.

The most common application of recombinant DNA is in basic research, in which the technology is important to most current work in the biological and biomedical sciences. Recombinant DNA is used to identify, map and sequence genes, and to determine their function. rDNA probes are employed in analyzing gene expression within individual cells, and throughout the tissues of whole organisms. Recombinant proteins are widely used as reagents in laboratory experiments and to generate antibody probes for examining protein synthesis within cells and organisms.

Many additional practical applications of recombinant DNA are found in industry, food production, human and veterinary medicine, agriculture, and bioengineering. Some specific examples are identified below.

Recombinant chymosin

> Found in rennet, chymosin is an enzyme required to manufacture cheese. It was the first genetically engineered food additive used commercially. Traditionally, processors obtained chymosin from rennet, a preparation

derived from the fourth stomach of milk-fed calves. Scientists engineered a non-pathogenic strain (K-12) of *E. coli* bacteria for large-scale laboratory production of the enzyme. This microbiologically produced recombinant enzyme, identical structurally to the calf derived enzyme, costs less and is produced in abundant quantities. Today about 60% of U.S. hard cheese is made with genetically engineered chymosin. In 1990, FDA granted chymosin "generally recognized as safe" (GRAS) status based on data showing that the enzyme was safe.

Recombinant human insulin

Almost completely replaced insulin obtained from animal sources (e.g. pigs and cattle) for the treatment of insulin-dependent diabetes. A variety of different recombinant insulin preparations are in widespread use. Recombinant insulin is synthesized by inserting the human insulin gene into *E. coli*, or yeast (saccharomyces cerevisiae) which then produces insulin for human use.

Recombinant human growth hormone (HGH, somatotropin)

Administered to patients whose pituitary glands generate insufficient quantities to support normal growth and development. Before recombinant HGH became available, HGH for therapeutic use was obtained from pituitary glands of cadavers. This unsafe practice led to some patients developing Creutzfeldt–Jakob disease. Recombinant HGH eliminated this problem, and is now used therapeutically. It has also been misused as a performance-enhancing drug by athletes and others.

Recombinant blood clotting factor VIII

A blood-clotting protein that is administered to patients with forms of the bleeding disorder hemophilia, who are unable to produce factor VIII in quantities sufficient to support normal blood coagulation. Before the development of recombinant factor VIII, the protein was obtained by processing large quantities of human blood from multiple donors, which carried a very high risk of transmission of blood borne infectious diseases, for example HIV and hepatitis.

Recombinant hepatitis B vaccine

Hepatitis B infection is controlled through the use of a recombinant hepatitis B vaccine, which contains a form of the hepatitis B virus surface antigen that is produced in yeast cells. The development of the recombinant subunit vaccine was an important and necessary development because hepatitis B virus, unlike other common viruses such as polio virus, cannot be grown in vitro. Vaccine information from Hepatitis B Foundation.

Diagnosis of infection with HIV

> Each of the three widely used methods for diagnosing HIV infection has been developed using recombinant DNA. The antibody test (ELISA or western blot) uses a recombinant HIV protein to test for the presence of antibodies that the body has produced in response to an HIV infection. The DNA test looks for the presence of HIV genetic material using reverse transcription polymerase chain reaction (RT-PCR). Development of the RT-PCR test was made possible by the molecular cloning and sequence analysis of HIV genomes.

Golden rice

> A recombinant variety of rice that has been engineered to express the enzymes responsible for β-carotene biosynthesis. This variety of rice holds substantial promise for reducing the incidence of vitamin A deficiency in the world's population. Golden rice is not currently in use, pending the resolution of regulatory and intellectual property issues.

Herbicide-resistant crops

> Commercial varieties of important agricultural crops (including soy, maize/corn, sorghum, canola, alfalfa and cotton) have been developed that incorporate a recombinant gene that results in resistance to the herbicide glyphosate (trade name *Roundup*), and simplifies weed control by glyphosate application. These crops are in common commercial use in several countries.

Insect-resistant crops

> *Bacillus thuringeiensis* is a bacterium that naturally produces a protein (Bt toxin) with insecticidal properties. The bacterium has been applied to crops as an insect-control strategy for many years, and this practice has been widely adopted in agriculture and gardening. Recently, plants have been developed that express a recombinant form of the bacterial protein, which may effectively control some insect predators. Environmental issues associated with the use of these transgenic crops have not been fully resolved.

History

The idea of recombinant DNA was first proposed by Peter Lobban, a graduate student of Prof. Dale Kaiser in the Biochemistry Department at Stanford University Medical School. The first publications describing the successful production and intracellular replication of recombinant DNA appeared in 1972 and 1973. Stanford University applied for a US patent on recombinant DNA in 1974, listing the inventors as Stanley N. Cohen and Herbert W. Boyer; this patent was awarded in 1980. The first licensed drug generated using recombinant DNA technology was human insulin, developed by Genentech and Licensed by Eli Lilly and Company.

Controversy

Scientists associated with the initial development of recombinant DNA methods recognized that the potential existed for organisms containing recombinant DNA to have undesirable or dangerous properties. At the 1975 Asilomar Conference on Recombinant DNA, these concerns were discussed and a voluntary moratorium on recombinant DNA research was initiated for experiments that were considered particularly risky. This moratorium was widely observed until the National Institutes of Health (USA) developed and issued formal guidelines for rDNA work. Today, recombinant DNA molecules and recombinant proteins are usually not regarded as dangerous. However, concerns remain about some organisms that express recombinant DNA, particularly when they leave the laboratory and are introduced into the environment or food chain.

Cloning

Many organisms, including aspen trees, reproduce by cloning.

In biology, cloning is the process of producing similar populations of genetically identical individuals that occurs in nature when organisms such as bacteria, insects or plants reproduce asexually. Cloning in biotechnology refers to processes used to create copies of DNA (deoxyribonucleic acid) fragments (molecular cloning), cells (cell cloning), or organisms. The term also refers to the production of multiple copies of a product such as digital media or software.

The term clone referes to the process whereby a new plant can be created from a twig. In horticulture, the spelling *clon* was used until the twentieth century; the final *e* came into use to indicate the vowel is a "long o" instead of a "short o". Since the term entered the popular lexicon in a more general context, the spelling *clone* has been used exclusively.

In botany, the term lusus was traditionally used.

Natural Cloning

Cloning is a natural form of reproduction that has allowed life forms to spread for more than 50 thousand years. It is the reproduction method used by plants, fungi, and bacteria, and is also the way that clonal colonies reproduce themselves. Examples of these organisms include blueberry plants, hazel trees, the Pando trees, the Kentucky coffeetree, *Myrica*s, and the American sweetgum.

Molecular Cloning

Molecular cloning refers to the process of making multiple molecules. Cloning is commonly used to amplify DNA fragments containing whole genes, but it can also be used to amplify any DNA sequence such as promoters, non-coding sequences and randomly fragmented DNA. It is used in a wide array of biological experiments and practical applications ranging from genetic fingerprinting to large scale protein production. Occasionally, the term cloning is misleadingly used to refer to the identification of the chromosomal location of a gene associated with a particular phenotype of interest, such as in positional cloning. In practice, localization of the gene to a chromosome or genomic region does not necessarily enable one to isolate or amplify the relevant genomic sequence. To amplify any DNA sequence in a living organism, that sequence must be linked to an origin of replication, which is a sequence of DNA capable of directing the propagation of itself and any linked sequence. However, a number of other features are needed, and a variety of specialised cloning vectors (small piece of DNA into which a foreign DNA fragment can be inserted) exist that allow protein production, affinity tagging, single stranded RNA or DNA production and a host of other molecular biology tools.

Cloning of any DNA fragment essentially involves four steps

1. fragmentation - breaking apart a strand of DNA

2. ligation - gluing together pieces of DNA in a desired sequence

3. transfection - inserting the newly formed pieces of DNA into cells

4. screening/selection - selecting out the cells that were successfully transfected with the new DNA

Although these steps are invariable among cloning procedures a number of alternative routes can be selected; these are summarized as a *cloning strategy*.

Initially, the DNA of interest needs to be isolated to provide a DNA segment of suitable size. Subsequently, a ligation procedure is used where the amplified fragment is inserted into a vector (piece of DNA). The vector (which is frequently circular) is linearised using restriction enzymes, and incubated with the fragment of interest under appropriate conditions with an enzyme called DNA ligase. Following ligation the vector with the insert of interest is transfected into cells. A number of alternative techniques are available, such as chemical sensitivation of cells, electroporation, optical injection and biolistics. Finally, the transfected cells are cultured. As the aforementioned procedures are of particularly low efficiency, there is a need to identify the cells that have been successfully transfected with the vector construct containing the desired insertion sequence in the required orientation. Modern cloning vectors include selectable antibiotic resistance markers, which allow only cells in which the vector has been transfected, to grow. Additionally, the cloning vectors may contain colour selection markers, which provide blue/white screening (alpha-factor complementation) on X-gal medium. Nevertheless, these selection steps do not absolutely guarantee that the DNA insert is present in the cells obtained. Further investigation of the resulting colonies must be required to confirm that cloning was successful. This may be accomplished by means of PCR, restriction fragment analysis and/or DNA sequencing.

Cell Cloning

Cloning Unicellular Organisms

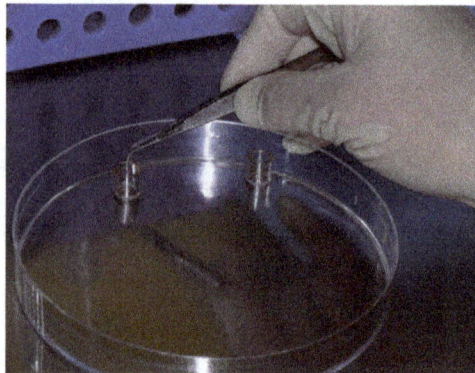

Cloning cell-line colonies using cloning rings

Cloning a cell means to derive a population of cells from a single cell. In the case of unicellular organisms such as bacteria and yeast, this process is remarkably simple and essentially only requires the inoculation of the appropriate medium. However, in the case of cell cultures from multi-cellular organisms, cell cloning is an arduous task as these cells will not readily grow in standard media.

A useful tissue culture technique used to clone distinct lineages of cell lines involves the use of cloning rings (cylinders). In this technique a single-cell suspension of cells that have been exposed to a mutagenic agent or drug used to drive selection is plated at high dilution to create isolated colonies, each arising from a single and potentially clonal distinct cell. At an early growth stage when colonies consist of only a few cells, sterile polystyrene rings (cloning rings), which have been dipped in grease, are placed over an individual colony and a small amount of trypsin is added. Cloned cells are collected from inside the ring and transferred to a new vessel for further growth.

Cloning Stem Cells

Somatic-cell nuclear transfer, known as SCNT, can also be used to create embryos for research or therapeutic purposes. The most likely purpose for this is to produce embryos for use in stem cell research. This process is also called "research cloning" or "therapeutic cloning." The goal is not to create cloned human beings (called "reproductive cloning"), but rather to harvest stem cells that can be used to study human development and to potentially treat disease. While a clonal human blastocyst has been created, stem cell lines are yet to be isolated from a clonal source.

Therapeutic cloning is achieved by creating embryonic stem cells in the hopes of treating diseases such as diabetes and Alzheimer's. The process begins by removing the nucleus (containing the DNA) from an egg cell and inserting a nucleus from the adult cell to be cloned. In the case of someone with Alzheimer's disease, the nucleus from a skin cell of that patient is placed into an empty egg. The reprogrammed cell begins to develop into an embryo because the egg reacts with the transferred nucleus. The embryo will become genetically identical to the patient. The embryo will then form a blastocyst which has the potential to form/become any cell in the body.

The reason why SCNT is used for cloning is because somatic cells can be easily acquired and cultured in the lab. This process can either add or delete specific genomes of farm animals. A key point to remember is that cloning is achieved when the oocyte maintains its normal functions and instead of using sperm and egg genomes to replicate, the oocyte is inserted into the donor's somatic cell nucleus. The oocyte will react on the somatic cell nucleus, the same way it would on sperm cells.

The process of cloning a particular farm animal using SCNT is relatively the same for all animals. The first step is to collect the somatic cells from the animal that will be cloned. The somatic cells could be used immediately or stored in the laboratory for later use. The hardest part of SCNT is removing maternal DNA from an oocyte at metaphase II. Once this has been done, the somatic nucleus can be inserted into an egg cytoplasm. This creates a one-cell embryo. The grouped somatic cell and egg cytoplasm are then introduced to an electrical current. This energy will hopefully allow the cloned embryo to begin development. The successfully developed embryos are then placed in surrogate recipients, such as a cow or sheep in the case of farm animals.

SCNT is seen as a good method for producing agriculture animals for food consumption. It successfully cloned sheep, cattle, goats, and pigs. Another benefit is SCNT is seen as a solution to clone endangered species that are on the verge of going extinct. However, stresses placed on both the egg cell and the introduced nucleus can be enormous, which led to a high loss in resulting cells in early research. For example, the cloned sheep Dolly was born after 277 eggs were used for SCNT, which created 29 viable embryos. Only three of these embryos survived until birth, and only one survived to adulthood. As the procedure could not be automated, and had to be performed manually under a microscope, SCNT was very resource intensive. The biochemistry involved in reprogramming the differentiated somatic cell nucleus and activating the recipient egg was also far from being well-understood. However, by 2014 researchers were reporting cloning success rates of seven to eight out of ten and in 2016, a Korean Company Sooam Biotech was reported to be producing 500 cloned embryos per day.

In SCNT, not all of the donor cell's genetic information is transferred, as the donor cell's mitochondria that contain their own mitochondrial DNA are left behind. The resulting hybrid cells retain those mitochondrial structures which originally belonged to the egg. As a consequence, clones such as Dolly that are born from SCNT are not perfect copies of the donor of the nucleus.

Organism Cloning

Organism cloning (also called reproductive cloning) refers to the procedure of creating a new multicellular organism, genetically identical to another. In essence this form of cloning is an asexual method of reproduction, where fertilization or inter-gamete contact does not take place. Asexual reproduction is a naturally occurring phenomenon in many species, including most plants and some insects. Scientists have made some major achievements with cloning, including the asexual reproduction of sheep and cows. There is a lot of ethical debate over whether or not cloning should be used. However, cloning, or asexual propagation, has been common practice in the horticultural world for hundreds of years.

Horticultural

The term *clone* is used in horticulture to refer to descendants of a single plant which were produced by vegetative reproduction or apomixis. Many horticultural plant cultivars are clones, having been derived from a single individual, multiplied by some process other than sexual reproduction. As an example, some European cultivars of grapes represent clones that have been propagated for over two millennia. Other examples are potato and banana. Grafting can be regarded as cloning, since all the shoots and branches coming from the graft are genetically a clone of a single individual, but this particular kind of cloning has not come under ethical scrutiny and is generally treated as an entirely different kind of operation.

Many trees, shrubs, vines, ferns and other herbaceous perennials form clonal colonies naturally. Parts of an individual plant may become detached by fragmentation and grow on to become separate clonal individuals. A common example is in the vegetative reproduction of moss and liverwort gametophyte clones by means of gemmae. Some vascular plants e.g. dandelion and certain viviparous grasses also form seeds asexually, termed apomixis, resulting in clonal populations of genetically identical individuals.

Parthenogenesis

Clonal derivation exists in nature in some animal species and is referred to as parthenogenesis (reproduction of an organism by itself without a mate). This is an asexual form of reproduction that is only found in females of some insects, crustaceans, nematodes, fish (for example the hammerhead shark), the Komodo dragon and lizards. The growth and development occurs without fertilization by a male. In plants, parthenogenesis means the development of an embryo from an unfertilized egg cell, and is a component process of apomixis. In species that use the XY sex-determination system, the offspring will always be female. An example is the little fire ant (*Wasmannia auropunctata*), which is native to Central and South America but has spread throughout many tropical environments.

Artificial Cloning of Organisms

Artificial cloning of organisms may also be called *reproductive cloning*.

First Moves

Hans Spemann, a German embryologist was awarded a Nobel Prize in Physiology or Medicine in 1935 for his discovery of the effect now known as embryonic induction, exercised by various parts of the embryo, that directs the development of groups of cells into particular tissues and organs. In 1928 he and his student, Hilde Mangold, were the first to perform somatic-cell nuclear transfer using amphibian embryos – one of the first moves towards cloning.

Methods

Reproductive cloning generally uses "somatic cell nuclear transfer" (SCNT) to create animals that are genetically identical. This process entails the transfer of a nucleus from a donor adult cell (somatic cell) to an egg from which the nucleus has been removed, or to a cell from a blastocyst from which the nucleus has been removed. If the egg begins to divide normally it is transferred into the uterus of the surrogate mother. Such clones are not strictly identical since the somatic cells may contain mutations in their nuclear DNA. Additionally, the mitochondria in the cytoplasm also contains DNA and during SCNT this mitochondrial DNA is wholly from the cytoplasmic donor's egg, thus the mitochondrial genome is not the same as that of the nucleus donor cell from which it was produced. This may have important implications for cross-species nuclear transfer in which nuclear-mitochondrial incompatibilities may lead to death.

Artificial *embryo splitting* or *embryo twinning*, a technique that creates monozygotic twins from a single embryo, is not considered in the same fashion as other methods of cloning. During that procedure, a donor embryo is split in two distinct embryos, that can then be transferred via embryo transfer. It is optimally performed at the 6- to 8-cell stage, where it can be used as an expansion of IVF to increase the number of available embryos. If both embryos are successful, it gives rise to monozygotic (identical) twins.

Dolly the Sheep

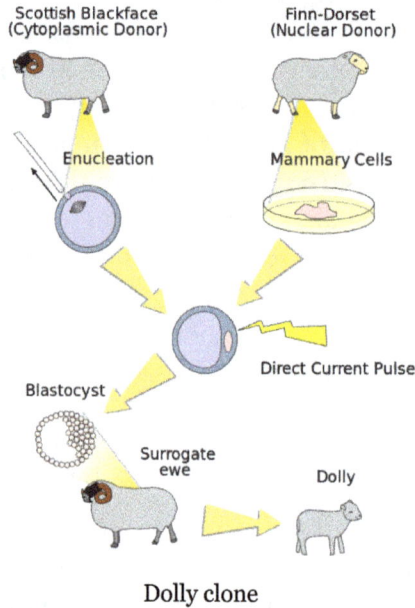

Dolly clone

Dolly, a Finn-Dorset ewe, was the first mammal to have been successfully cloned from an adult somatic cell. Dolly was formed by taking a cell from the udder of her 6-year old biological mother. Dolly's embryo was created by taking the cell and inserting it into a sheep ovum. It took 434 attempts before an embryo was successful. The embryo was then placed inside a female sheep that went through a normal pregnancy. She was cloned at the Roslin Institute in Scotland by British scientists Sir Ian Wilmut and Keith Campbell and lived there from her birth in 1996 until her death in 2003 when she was six. She was born on 5 July 1996 but not announced to the world until 22 February 1997. Her stuffed remains were placed at Edinburgh's Royal Museum, part of the National Museums of Scotland.

Dolly was publicly significant because the effort showed that genetic material from a specific adult cell, programmed to express only a distinct subset of its genes, can be reprogrammed to grow an entirely new organism. Before this demonstration, it had been shown by John Gurdon that nuclei from differentiated cells could give rise to an entire organism after transplantation into an enucleated egg. However, this concept was not yet demonstrated in a mammalian system.

The first mammalian cloning (resulting in Dolly the sheep) had a success rate of 29 embryos per 277 fertilized eggs, which produced three lambs at birth, one of which lived. In a bovine experiment involving 70 cloned calves, one-third of the calves died young. The first successfully cloned horse, Prometea, took 814 attempts. Notably, although the first clones were frogs, no adult cloned frog has yet been produced from a somatic adult nucleus donor cell.

There were early claims that Dolly the sheep had pathologies resembling accelerated aging. Scientists speculated that Dolly's death in 2003 was related to the shortening of telomeres, DNA-protein complexes that protect the end of linear chromosomes. However, other researchers, including Ian Wilmut who led the team that successfully cloned Dolly, argue that Dolly's early death due to respiratory infection was unrelated to deficiencies with the cloning process. This idea that the nuclei have not irreversibly aged was shown in 2013 to be true for mice.

Dolly was named after performer Dolly Parton because the cells cloned to make her were from a mammary gland cell, and Parton is known for her ample cleavage.

Species Cloned

The modern cloning techniques involving nuclear transfer have been successfully performed on several species. Notable experiments include:

- Tadpole: (1952) Robert Briggs and Thomas J. King had successfully cloned northern leopard frogs: thirty-five complete embryos and twenty-seven tadpoles from one-hundred and four successful nuclear transfers.

- Carp: (1963) In China, embryologist Tong Dizhou produced the world's first cloned fish by inserting the DNA from a cell of a male carp into an egg from a female carp.

- Mice: (1986) A mouse was successfully cloned from an early embryonic cell. Soviet scientists Chaylakhyan, Veprencev, Sviridova, and Nikitin had the mouse "Masha" cloned. Research was published in the magazine "Biofizika" volume XXXII, issue 5 of 1987.

- Sheep: Marked the first mammal being cloned (1984) from early embryonic cells by Steen Willadsen. Megan and Morag cloned from differentiated embryonic cells in June 1995 and Dolly the sheep from a somatic cell in 1996.

- Rhesus monkey: Tetra (January 2000) from embryo splitting

- Pig: the first cloned pigs (March 2000). By 2014, BGI in China was producing 500 cloned pigs a year to test new medicines.

- Gaur: (2001) was the first endangered species cloned.

- Cattle: Alpha and Beta (males, 2001) and (2005) Brazil

- Cat: CopyCat "CC" (female, late 2001), Little Nicky, 2004, was the first cat cloned for commercial reasons

- Rat: Ralph, the first cloned rat (2003)

- Mule: Idaho Gem, a john mule born 4 May 2003, was the first horse-family clone.

- Horse: Prometea, a Haflinger female born 28 May 2003, was the first horse clone.

- Dog: Snuppy, a male Afghan hound was the first cloned dog (2005).

- Wolf: Snuwolf and Snuwolffy, the first two cloned female wolves (2005).

- Water buffalo: Samrupa was the first cloned water buffalo. It was born on 6 February 2009, at India's Karnal National Diary Research Institute but died five days later due to lung infection.

- Pyrenean ibex (2009) was the first extinct animal to be cloned back to life; the clone lived for seven minutes before dying of lung defects.

- Camel: (2009) Injaz, is the first cloned camel.

- Pashmina goat: (2012) Noori, is the first cloned pashmina goat. Scientists at the faculty of veterinary sciences and animal husbandry of Sher-e-Kashmir University of Agricultural Sciences and Technology of Kashmir successfully cloned the first Pashmina goat (Noori) using the advanced reproductive techniques under the leadership of Riaz Ahmad Shah.

- Gastric brooding frog: (2013) The gastric brooding frog, *Rheobatrachus silus*, thought to have been extinct since 1983 was cloned in Australia, although the embryos died after a few days.

Human Cloning

Human cloning is the creation of a genetically identical copy of a human. The term is generally used to refer to artificial human cloning, which is the reproduction of human cells and tissues. It does not refer to the natural conception and delivery of identical twins. The possibility of human cloning has raised controversies. These ethical concerns have prompted several nations to pass legislature regarding human cloning and its legality.

Two commonly discussed types of theoretical human cloning are *therapeutic cloning* and *reproductive cloning*. Therapeutic cloning would involve cloning cells from a human for use in medicine and transplants, and is an active area of research, but is not in medical practice anywhere in the world, as of 2014. Two common methods of therapeutic cloning that are being researched are somatic-cell nuclear transfer and, more recently, pluripotent stem cell induction. Reproductive cloning would involve making an entire cloned human, instead of just specific cells or tissues.

Ethical Issues of Cloning

There are a variety of ethical positions regarding the possibilities of cloning, especially human cloning. While many of these views are religious in origin, the questions raised by cloning are faced by secular perspectives as well. Perspectives on human cloning are theoretical, as human therapeutic and reproductive cloning are not commercially used; animals are currently cloned in laboratories and in livestock production.

Advocates support development of therapeutic cloning in order to generate tissues and whole organs to treat patients who otherwise cannot obtain transplants, to avoid the need for immunosuppressive drugs, and to stave off the effects of aging. Advocates for reproductive cloning believe that parents who cannot otherwise procreate should have access to the technology.

Opponents of cloning have concerns that technology is not yet developed enough to be safe and that it could be prone to abuse (leading to the generation of humans from whom organs and tissues would be harvested), as well as concerns about how cloned individuals could integrate with families and with society at large.

Religious groups are divided, with some opposing the technology as usurping "God's place" and, to the extent embryos are used, destroying a human life; others support therapeutic cloning's potential life-saving benefits.

Cloning of animals is opposed by animal-groups due to the number of cloned animals that suffer from malformations before they die, and while food from cloned animals has been approved by the US FDA, its use is opposed by groups concerned about food safety.

Cloning Extinct and Endangered Species

Cloning, or more precisely, the reconstruction of functional DNA from extinct species has, for decades, been a dream. Possible implications of this were dramatized in the 1984 novel *Carnosaur* and the 1990 novel *Jurassic Park*. The best current cloning techniques have an average success rate of 9.4 percent (and as high as 25 percent) when working with familiar species such as mice, while cloning wild animals is usually less than 1 percent successful. Several tissue banks have come into existence, including the "Frozen Zoo" at the San Diego Zoo, to store frozen tissue from the world's rarest and most endangered species.

In 2001, a cow named Bessie gave birth to a cloned Asian gaur, an endangered species, but the calf died after two days. In 2003, a banteng was successfully cloned, followed by three African wildcats from a thawed frozen embryo. These successes provided hope that similar techniques (using surrogate mothers of another species) might be used to clone extinct species. Anticipating this possibility, tissue samples from the last *bucardo* (Pyrenean ibex) were frozen in liquid nitrogen immediately after it died in 2000. Researchers are also considering cloning endangered species such as the giant panda and cheetah.

In 2002, geneticists at the Australian Museum announced that they had replicated DNA of the thylacine (Tasmanian tiger), at the time extinct for about 65 years, using polymerase chain reaction. However, on 15 February 2005 the museum announced that it was stopping the project after tests showed the specimens' DNA had been too badly degraded by the (ethanol) preservative. On 15 May 2005 it was announced that the thylacine project would be revived, with new participation from researchers in New South Wales and Victoria.

In January 2009, for the first time, an extinct animal, the Pyrenean ibex mentioned above was cloned, at the Centre of Food Technology and Research of Aragon, using the preserved frozen cell nucleus of the skin samples from 2001 and domestic goat egg-cells. The ibex died shortly after birth due to physical defects in its lungs.

One of the most anticipated targets for cloning was once the woolly mammoth, but attempts to extract DNA from frozen mammoths have been unsuccessful, though a joint Russo-Japanese team is currently working toward this goal. In January 2011, it was reported by Yomiuri Shimbun that a team of scientists headed by Akira Iritani of Kyoto University had built upon research by Dr. Wakayama, saying that they will extract DNA from a mammoth carcass that had been preserved in a Russian laboratory and insert it into the egg cells of an African elephant in hopes of producing a mammoth embryo. The researchers said they hoped to produce a baby mammoth within six years. It was noted, however that the result, if possible, would be an elephant-mammoth hybrid rather than a true mammoth. Another problem is the survival of the reconstructed mammoth: ruminants rely on a symbiosis with specific microbiota in their stomachs for digestion.

Scientists at the University of Newcastle and University of New South Wales announced in March 2013 that the very recently extinct gastric-brooding frog would be the subject of a cloning attempt to resurrect the species.

Many such "de-extinction" projects are described in the Long Now Foundation's Revive and Restore Project.

Lifespan

After an eight-year project involving the use of a pioneering cloning technique, Japanese researchers created 25 generations of healthy cloned mice with normal lifespans, demonstrating that clones are not intrinsically shorter-lived than naturally born animals. Other sources have noted that the offspring of clones tend to be healthier than the original clones and indistinguishable from animals produced naturally.

In a detailed study released in 2016 and less detailed studies by others suggest that once cloned animals get past the first month or two of life they are generally healthy. However, early pregnancy loss and neonatal losses are still greater with cloning than natural conception or assisted reproduction (IVF). Current research endeavors are attempting to overcome this problem.

In Popular Culture

In an article in the 8 November 1993 article of *Time*, cloning was portrayed in a negative way, modifying Michelangelo's *Creation of Adam* to depict Adam with five identical hands. *Newsweek's* 10 March 1997 issue also critiqued the ethics of human cloning, and included a graphic depicting identical babies in beakers.

Cloning is a recurring theme in a wide variety of contemporary science fiction, ranging from action films such as *Jurassic Park* (1993), *The 6th Day* (2000), *Resident Evil* (2002), *Star Wars: Episode II* (2002) and *The Island* (2005), to comedies such as Woody Allen's 1973 film *Sleeper*.

Science fiction has used cloning, most commonly and specifically human cloning, due to the fact that it brings up controversial questions of identity. *A Number* is a 2002 play by English playwright Caryl Churchill which addresses the subject of human cloning and identity, especially nature and nurture. The story, set in the near future, is structured around the conflict between a father (Salter) and his sons (Bernard 1, Bernard 2, and Michael Black) – two of whom are clones of the first one. *A Number* was adapted by Caryl Churchill for television, in a co-production between the BBC and HBO Films.

A recurring sub-theme of cloning fiction is the use of clones as a supply of organs for transplantation. The 2005 Kazuo Ishiguro novel *Never Let Me Go* and the 2010 film adaption are set in an alternate history in which cloned humans are created for the sole purpose of providing organ donations to naturally born humans, despite the fact that they are fully sentient and self-aware. The 2005 film *The Island* revolves around a similar plot, with the exception that the clones are unaware of the reason for their existence.

The use of human cloning for military purposes has also been explored in several works. *Star Wars* portrays human cloning in *Clone Wars*.

The exploitation of human clones for dangerous and undesirable work was examined in the 2009 British science fiction film *Moon*. In the futuristic novel *Cloud Atlas* and subsequent film, one of the story lines focuses on a genetically-engineered fabricant clone named Sonmi~451 who is one of millions raised in an artificial "wombtank," destined to serve from birth. She is one of thousands of clones created for manual and emotional labor; Sonmi herself works as a server in a restaurant. She later discovers that the sole source of food for clones, called 'Soap', is manufactured from the clones themselves.

Cloning has been used in fiction as a way of recreating historical figures. In the 1976 Ira Levin novel *The Boys from Brazil* and its 1978 film adaptation, Josef Mengele uses cloning to create copies of Adolf Hitler.

In 2012, a Japanese television show named "Bunshin" was created. The story's main character, Mariko, is a woman studying child welfare in Hokkaido. She grew up always doubtful about the love from her mother, who looked nothing like her and who died nine years before. One day, she finds some of her mother's belongings at a relative's house, and heads to Tokyo to seek out the truth behind her birth. She later discovered that she was a clone.

In the 2013 television show *Orphan Black*, cloning is used as a scientific study on the behavioral adaptation of the clones. In a similar vein, the book *The Double* by Nobel Prize winner José Saramago explores the emotional experience of a man who discovers that he is a clone.

Restriction Enzyme

A restriction enzyme or restriction endonuclease is an enzyme that cuts DNA at or near specific recognition nucleotide sequences known as restriction sites. Restriction enzymes are commonly classified into four types, which differ in their structure and whether they cut their DNA substrate at their recognition site, or if the recognition and cleavage sites are separate from one another. To cut DNA, all restriction enzymes make two incisions, once through each sugar-phosphate backbone (i.e. each strand) of the DNA double helix.

These enzymes are found in bacteria and archaea and provide a defense mechanism against invading viruses. Inside a prokaryote, the restriction enzymes selectively cut up *foreign* DNA in a process called *restriction*; meanwhile, host DNA is protected by a modification enzyme (a methyltransferase) that modifies the prokaryotic DNA and blocks cleavage. Together, these two processes form the restriction modification system.

Over 3000 restriction enzymes have been studied in detail, and more than 600 of these are available commercially. These enzymes are routinely used for DNA modification in laboratories, and are a vital tool in molecular cloning.

History

The term restriction enzyme originated from the studies of phage λ and the phenomenon of host-controlled restriction and modification of a bacterial virus. The phenomenon was first identified in work done in the laboratories of Salvador Luria and Giuseppe Bertani in early 1950s. It was found that, for a bacteriophage λ that can grow well in one strain of *Escherichia coli*, for example *E. coli* C, when grown in another strain, for example *E. coli* K, its yields can drop significantly, by as much as 3-5 orders of magnitude. The host cell, in this example *E. coli* K, is known as the restricting host and appears to have the ability to reduce the biological activity of the phage λ. If a phage becomes established in one strain, the ability of that phage to grow also becomes restricted in other strains. In the 1960s, it was shown in work done in the laboratories of Werner Arber and Matthew Meselson that the restriction is caused by an enzymatic cleavage of the phage DNA, and the enzyme involved was therefore termed a restriction enzyme.

The restriction enzymes studied by Arber and Meselson were type I restriction enzymes, which cleave DNA randomly away from the recognition site. In 1970, Hamilton O. Smith, Thomas Kelly and Kent Wilcox isolated and characterized the first type II restriction enzyme, *Hind*II, from the bacterium *Haemophilus influenzae*. Restriction enzymes of this type are more useful for laboratory work as they cleave DNA at the site of their recognition sequence. Later, Daniel Nathans and Kathleen Danna showed that cleavage of simian virus 40 (SV40) DNA by restriction enzymes yields specific fragments that can be separated using polyacrylamide gel electrophoresis, thus showing that restriction enzymes can also be used for mapping DNA. For their work in the discovery and characterization of restriction enzymes, the 1978 Nobel Prize for Physiology or Medicine was awarded to Werner Arber, Daniel Nathans, and Hamilton O. Smith. The discovery of restriction enzymes allows DNA to be manipulated, leading to the development of recombinant DNA technology that has many applications, for example, allowing the large scale production of proteins such as human insulin used by diabetics.

Restriction enzymes likely evolved from a common ancestor and became widespread via horizontal gene transfer. In addition, there is mounting evidence that restriction endonucleases evolved as a selfish genetic element.

Recognition Site

$$5'...GAT\ |\ ATC...3'$$
$$3'...CTA\ |\ TAG...5'$$

A palindromic recognition site reads the same on the reverse strand as it does on the forward strand when both are read in the same orientation

Restriction enzymes recognize a specific sequence of nucleotides and produce a double-stranded cut in the DNA. The recognition sequences can also be classified by the number of bases in its recognition site, usually between 4 and 8 bases, and the amount of bases in the sequence will determine how often the site will appear by chance in any given genome, e.g., a 4-base pair sequence would theoretically occur once every 4^4 or 256bp, 6 bases, 4^6 or 4,096bp, and 8 bases would be 4^8 or 65,536bp. Many of them are palindromic, meaning the base sequence reads the same backwards and forwards. In theory, there are two types of palindromic sequences that can be possible in DNA. The *mirror-like* palindrome is similar to those found in ordinary text, in which a sequence reads the same forward and backward on a single strand of DNA, as in GTAATG. The *inverted repeat* palindrome is also a sequence that reads the same forward and backward, but the forward and backward sequences are found in complementary DNA strands (i.e., of double-stranded DNA), as in GTATAC (GTATAC being complementary to CATATG). Inverted repeat palindromes are more common and have greater biological importance than mirror-like palindromes.

*Eco*RI digestion produces "sticky" ends,

$$
\begin{array}{l}
\text{G\,A\,A\,T\,T\,C} \\
\text{C\,T\,T\,A\,A\,G}
\end{array}
$$

whereas SmaI restriction enzyme cleavage produces "blunt" ends:

$$
\begin{array}{l}
\text{C\,C\,C\,G\,G\,G} \\
\text{G\,G\,G\,C\,C\,C}
\end{array}
$$

Recognition sequences in DNA differ for each restriction enzyme, producing differences in the length, sequence and strand orientation (5' end or 3' end) of a sticky-end "overhang" of an enzyme restriction.

Different restriction enzymes that recognize the same sequence are known as neoschizomers. These often cleave in different locales of the sequence. Different enzymes that recognize and cleave in the same location are known as isoschizomers.

Types

Naturally occurring restriction endonucleases are categorized into four groups (Types I, II III, and IV) based on their composition and enzyme cofactor requirements, the nature of their target sequence, and the position of their DNA cleavage site relative to the target sequence. DNA sequence analyses of restriction enzymes however show great variations, indicating that there are more than four types. All types of enzymes recognize specific short DNA sequences and carry out the endonucleolytic cleavage of DNA to give specific fragments with terminal 5'-phosphates. They differ in their recognition sequence, subunit composition, cleavage position, and cofactor requirements, as summarised below:

- Type I enzymes (EC 3.1.21.3) cleave at sites remote from a recognition site; require both ATP and S-adenosyl-L-methionine to function; multifunctional protein with both restriction and methylase (EC 2.1.1.72) activities.

- Type II enzymes (EC 3.1.21.4) cleave within or at short specific distances from a recognition site; most require magnesium; single function (restriction) enzymes independent of methylase.

- Type III enzymes (EC 3.1.21.5) cleave at sites a short distance from a recognition site; require ATP (but do not hydrolyse it); S-adenosyl-L-methionine stimulates the reaction but is not required; exist as part of a complex with a modification methylase (EC 2.1.1.72).

- Type IV enzymes target modified DNA, e.g. methylated, hydroxymethylated and glucosyl-hydroxymethylated DNA

Type I

Type I restriction enzymes were the first to be identified and were first identified in two different strains (K-12 and B) of *E. coli*. These enzymes cut at a site that differs, and is a random distance (at least 1000 bp) away, from their recognition site. Cleavage at these random sites follows a process of DNA translocation, which shows that these enzymes are also molecular motors. The recognition site is asymmetrical and is composed of two specific portions—one containing 3–4 nucleotides, and another containing 4–5 nucleotides—separated by a non-specific spacer of about 6–8 nucleotides. These enzymes are multifunctional and are capable of both restriction and modification activities, depending upon the methylation status of the target DNA. The cofactors S-Adenosyl methionine (AdoMet), hydrolyzed adenosine triphosphate (ATP), and magnesium (Mg^{2+}) ions, are required for their full activity. Type I restriction enzymes possess three subunits called HsdR, HsdM, and HsdS; HsdR is required for restriction; HsdM is necessary for adding methyl groups to host DNA (methyltransferase activity), and HsdS is important for specificity of the recognition (DNA-binding) site in addition to both restriction (DNA cleavage) and modification (DNA methyltransferase) activity.

Type II

Typical type II restriction enzymes differ from type I restriction enzymes in several ways. They form homodimers, with recognition sites that are usually undivided and palindromic and 4–8 nucleotides in length. They recognize and cleave DNA at the same site, and they do not use ATP or AdoMet for their activity—they usually require only Mg^{2+} as a cofactor. These are the most commonly available and used restriction enzymes. In the 1990s and early 2000s, new enzymes from this family were discovered that did not follow all the classical criteria of this enzyme class, and new subfamily nomenclature was developed to divide this large family into subcategories based on deviations from typical characteristics of type II enzymes. These subgroups are defined using a letter suffix.

Type IIB restriction enzymes (e.g., BcgI and BplI) are multimers, containing more than one subunit. They cleave DNA on both sides of their recognition to cut out the recognition site. They require both AdoMet and Mg^{2+} cofactors. Type IIE restriction endonucleases (e.g., NaeI) cleave DNA following interaction with two copies of their recognition sequence. One recognition site acts as the target for cleavage, while the other acts as an allosteric effector that speeds up or improves the efficiency of enzyme cleavage. Similar to type IIE enzymes, type IIF restriction endonucleases (e.g. NgoMIV) interact with two copies of their recognition sequence but cleave both sequences at the same time. Type IIG restriction endonucleases (e.g., Eco57I) do have a single subunit, like classical Type II restriction enzymes, but require the cofactor AdoMet to be active. Type IIM restriction endonucleases, such as DpnI, are able to recognize and cut methylated DNA. Type IIS restriction endonucleases (e.g., *Fok*I) cleave DNA at a defined distance from their non-palindromic asymmetric recognition sites; this characteristic is widely used to perform in-vitro cloning techniques such as Golden Gate cloning. These enzymes may function as dimers. Similarly, Type IIT restriction enzymes (e.g., Bpu10I and BslI) are composed of two different subunits. Some recognize palindromic sequences while others have asymmetric recognition sites.

Type III

Type III restriction enzymes (e.g., EcoP15) recognize two separate non-palindromic sequences that are inversely oriented. They cut DNA about 20–30 base pairs after the recognition site. These enzymes contain more than one subunit and require AdoMet and ATP cofactors for their roles in DNA methylation and restriction, respectively. They are components of prokaryotic DNA restriction-modification mechanisms that protect the organism against invading foreign DNA. Type III enzymes are hetero-oligomeric, multifunctional proteins composed of two subunits, Res and Mod. The Mod subunit recognises the DNA sequence specific for the system and is a modification methyltransferase; as such, it is functionally equivalent to the M and S subunits of type I restriction endonuclease. Res is required for restriction, although it has no enzymatic activity on its own. Type III enzymes recognise short 5–6 bp-long asymmetric DNA sequences and cleave 25–27 bp downstream to leave short, single-stranded 5' protrusions. They require the presence of two inversely oriented unmethylated recognition sites for restriction to occur. These enzymes methylate only one strand of the DNA, at the N-6 position of adenosyl residues, so newly replicated DNA will have only one strand methylated, which is sufficient to protect against restriction. Type III enzymes belong to the beta-subfamily of N6 adenine methyltransferases, containing the nine motifs that characterise this family, including motif I, the AdoMet binding pocket (FXGXG), and motif IV, the catalytic region (S/D/N (PP) Y/F).

Type IV

Type IV enzymes recognize modified, typically methylated DNA and are exemplified by the McrBC and Mrr systems of *E. coli*.

Type V

Type V restriction enzymes (e.g., the cas9-gRNA complex from CRISPRs) utilize guide RNAs to target specific non-palindromic sequences found on invading organisms. They can cut DNA of variable length, provided that a suitable guide RNA is provided. The flexibility and ease of use of these enzymes make them promising for future genetic engineering applications.

Artificial Restriction Enzymes

Artificial restriction enzymes can be generated by fusing a natural or engineered DNA binding domain to a nuclease domain (often the cleavage domain of the type IIS restriction enzyme *Fok*I). Such artificial restriction enzymes can target large DNA sites (up to 36 bp) and can be engineered to bind to desired DNA sequences. Zinc finger nucleases are the most commonly used artificial restriction enzymes and are generally used in genetic engineering applications, but can also be used for more standard gene cloning applications. Other artificial restriction enzymes are based on the DNA binding domain of TAL effectors.

In 2013, a new technology CRISPR-Cas9, based on a prokaryotic viral defense system, was engineered for editing the genome, and it was quickly adopted in laboratories.

In 2017 a group in Illinois announced using an Argonaute protein taken from Pyrococcus furiosus (PfAgo) along with guide DNA to edit DNA as artificial restriction enzymes.

Nomenclature

Derivation of the *Eco*RI name		
Abbreviation	**Meaning**	**Description**
E	*Escherichia*	genus
co	*coli*	specific epithet
R	RY13	strain
I	First identified	order of identification in the bacterium

Since their discovery in the 1970s, many restriction enzymes have been identified; for example, more than 3500 different Type II restriction enzymes have been characterized. Each enzyme is named after the bacterium from which it was isolated, using a naming system based on bacterial genus, species and strain. For example, the name of the *Eco*RI restriction enzyme was derived as shown in the box.

Applications

Isolated restriction enzymes are used to manipulate DNA for different scientific applications.

They are used to assist insertion of genes into plasmid vectors during gene cloning and protein production experiments. For optimal use, plasmids that are commonly used for gene cloning are modified to include a short *polylinker* sequence (called the multiple cloning site, or MCS) rich in restriction enzyme recognition sequences. This allows flexibility when inserting gene fragments into the plasmid vector; restriction sites contained naturally within genes influence the choice of endonuclease for digesting the DNA, since it is necessary to avoid restriction of wanted DNA while intentionally cutting the ends of the DNA. To clone a gene fragment into a vector, both plasmid DNA and gene insert are typically cut with the same restriction enzymes, and then glued together with the assistance of an enzyme known as a DNA ligase.

Restriction enzymes can also be used to distinguish gene alleles by specifically recognizing single base changes in DNA known as single nucleotide polymorphisms (SNPs). This is however only possible if a SNP alters the restriction site present in the allele. In this method, the restriction enzyme can be used to genotype a DNA sample without the need for expensive gene sequencing. The sample is first digested with the restriction enzyme to generate DNA fragments, and then the different sized fragments separated by gel electrophoresis. In general, alleles with correct restriction sites will generate two visible bands of DNA on the gel, and those with altered restriction sites will not be cut and will generate only a single band. A DNA map by restriction digest can also be generated that can give the relative positions of the genes. The different lengths of DNA generated by restriction digest also produce a specific pattern of bands after gel electrophoresis, and can be used for DNA fingerprinting.

In a similar manner, restriction enzymes are used to digest genomic DNA for gene analysis by Southern blot. This technique allows researchers to identify how many copies (or paralogues) of a gene are present in the genome of one individual, or how many gene mutations (polymorphisms) have occurred within a population. The latter example is called restriction fragment length polymorphism (RFLP).

Artificial restriction enzymes created by linking the *Fok*I DNA cleavage domain with an array of DNA binding proteins or zinc finger arrays, denoted zinc finger nucleases (ZFN), are a powerful tool for host genome editing due to their enhanced sequence specificity. ZFN work in pairs, their dimerization being mediated in-situ through the *Fok*I domain. Each zinc finger array (ZFA) is capable of recognizing 9–12 base pairs, making for 18–24 for the pair. A 5–7 bp spacer between the cleavage sites further enhances the specificity of ZFN, making them a safe and more precise tool that can be applied in humans. A recent Phase I clinical trial of ZFN for the targeted abolition of the CCR5 co-receptor for HIV-1 has been undertaken.

Others have proposed using the bacteria R-M system as a model for devising human anti-viral gene or genomic vaccines and therapies since the RM system serves an innate defense-role in bacteria by restricting tropism by bacteriophages. There is research on REases and ZFN that can cleave the DNA of various human viruses, including HSV-2, high-risk HPVs and HIV-1, with the ultimate goal of inducing target mutagenesis and aberrations of human-infecting viruses. Interestingly, the human genome already contains remnants of retroviral genomes that have been inactivated and harnessed for self-gain. Indeed, the mechanisms for silencing active L1 genomic retroelements by the three prime repair exonuclease 1 (TREX1) and excision repair cross complementing 1(ERCC) appear to mimic the action of RM-systems in bacteria, and the non-homologous end-joining (NHEJ) that follows the use of ZFN without a repair template.

Polymerase Chain Reaction

Polymerase chain reaction (PCR) is used to amplify a DNA sequence to produce millions of copies. Kary Mullis discovered the PCR and got Nobel Prize in Chemistry in 1993 for his discovery. Since then, PCR has been used in various applications in medicine, animal science, plant science, food science etc. The different events to develop present day PCR is given in the table below.

Year	The gradual breakthrough from the discovery of DNA structure to the invention of modern PCR
1950	Discovery of mechanism of DNA Replication by Arthur Kornberg. He discovered the first DNA polymerases and other factors like helicase and primers.
1976	Isolation of thermostable DNA polymerase from *T.aquaticus*.
1983	Mullis synthesized DNA oligo probes for Sickle cell anemia mutation.
1983	Repeated thermal cycling was first used for small segment of cloned gene.
1984	Mullis and Tom White tried designed experiments to test PCR on genomic DNA but the amplified product was not visible in agarose gel.
1985	Patent was filed for PCR and its applications focusing on sickle cell anemia mutation.
1985	The use of thermostable DNA polymerase in PCR was started. Out of only two enzymes (Taq and Bst) known at that time, Taq was found more suitable for PCR.
1985	First announcement of PCR technique in Salt Lake City.
1985-1987	Development of instrument for PCR and its reagents.

Different events in development of PCR

Principle of the technique: The whole process of PCR involves three main events, Denaturation, Annealing and Elongation. A DNA fragment of interest is used as a template and a pair of primers which are short oligonucleotides complimentary to the both strands of the template DNA. The purpose of primer is to initiate the DNA synthesis in the direction of 5' to 3'. The number of amplified DNA or the amplicons increases exponentially per cycle thus one molecule of DNA gives rise to 2,4,8,16 and so forth. This continuous doubling is carried out by a specific enzyme called DNA polymerase which sits at the unfinished double stranded DNA created by template DNA and primer. For further extension of the DNA, the polymerase enzyme require supply of other DNA-building blocks such as the nucleotides consisting of four bases Adenine (A), Thymine (T), Cytosine (C) and Guanine (G). The template, primer, polymerase and four bases are the main components for polymerase chain reaction.

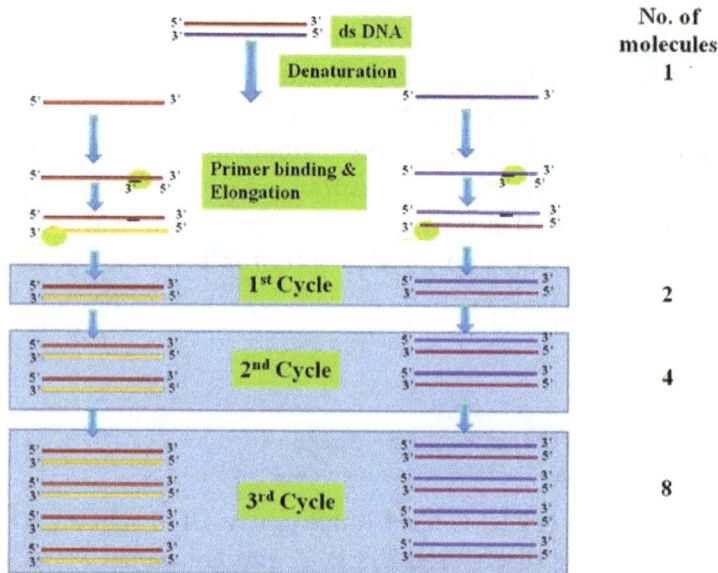

Basic Principle of polymerase chain reaction (PCR).

Methodology: PCR has three major events (Denaturation, Annealing and Elongation) to complete the amplification process. The complete process of PCR is as follows-

1. Initial denaturation: Heating the PCR mixture at 94° C to 96° C for 10min to ensure complete denaturation of template DNA. It is followed by the cyclic events which has different steps as described below:

    A. Denaturation: This is the first step in which the double stranded DNA template is denatured to form two single strand by heating at 95° C for 15-30 secs.

    B. Annealing: This is the annealing step where at lower temperature (usually 50-65° C) primers are allowed to bind to template DNA, annealing time is 15-30 secs and it depends on the length and bases of the primers. Generally annealing temperature is about 3-5° C below the melting temperature (T_m) of the pair of the primers is used.

    C. Elongation: This is the synthesis step where the polymerase perform synthesis of new strand in the 5' to 3' direction using primer and deoxyribonucleoside triphosphates (dNTPs). An average DNA polymerase adds about 1,000 bp/minute. Step 1,2,3 makes one cycle and in general 35-40 such cycles are performed in a typical PCR amplification.

2. After the cycles are completed, the reaction is held at 70-74° C for several minutes to allow final extension of the remaining DNA to be fully extended.

3. The reaction is complete and the resulting amplified nucleic acids are held at a low temperature (~4° C) until analysis.

Reagents: The reagents required for a complete PCR reaction volume is given in the table.

Reagents	Amount required
Template DNA	1pg-1ng for viral or short templates
	1ng-1µg for genomic DNA
Primers (forward and reverse primers)	0.1-0.5µM of each primer
Magnesium chloride	1.5-2.0 mM is optimal for *Taq* DNA polymerase
Deoxynucleotides (dNTPs)	Typical concentration is 200 µM of each dNTP
Taq DNA Polymerase	0.5–2.0 units per 50 µl reaction

Instrumentation: Thermal cycler is the instrument that carries out the amplification via polymerase chain reaction. Usually the three main events are repeated for 30-40 cycles to obtain detectable amount of product at the end of the cycles. The automated system performs the cyclic temperature changes required for enzymatic amplification of specific DNA segments in vitro using this PCR. The device has a thermal block with holes where tubes containing reaction mixtures can be inserted. The cycler varies the temperature of the block in discrete, pre-programmed steps using peltier effect.

Primers: A primer is a short oligonucleotide that serves as a starting point for DNA synthesis. In PCR, two primers are required to bind to each of the single stranded DNA (obtained after denaturation) flanking the target sequence. These are called Forward and Reverse primers. They primers are designed in such a way that they have a sequence complimentary to the sequence in the template DNA.

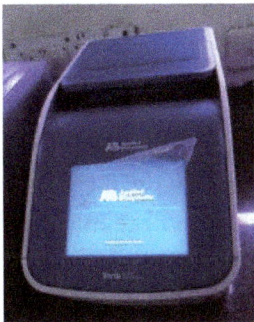

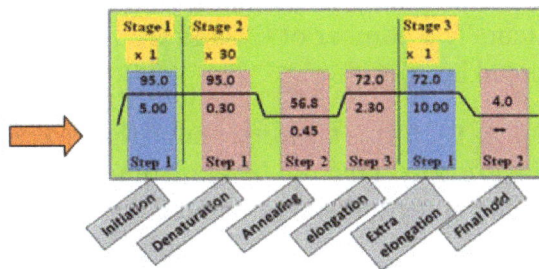

Representation of thermal cycler instrument showing the position of sample and schematic diagram of 30 cycle PCR.

Two restriction enzymes sites are added at the 5' end of each of the primer to facilitate cloning. The chosen restriction enzymes will not cut DNA fragment (non-cutters). Typically 3 to 4 nucleotides are added at the end of the restriction sites to allow efficient cutting by restriction enzymes.

Primer Designing and criteria: For a specific amplifications in PCR, good primer design is essential. The following parameters needs to be considered while designing a primer:

1. Primer length: Oligonucleotides between 18-24 bases is the ideal length which is long enough for adequate specificity and short enough for primers to bind easily to the template at the annealing temperature.

2. Primer melting temperature (T_m): Primers with melting temperatures in the range of 52-58°C generally gives the best results. The GC content of the sequence gives a fair indication of the primer T_m. The two primers should be prepared in such a way that their Tm difference should not be more than 2°C otherwise it will result in poor annealing efficiency. Tm can be calculated by the following formula:

FOR A PRIMER LENGTH <14 NUCLEO-TIDES	FOR A PRIMER LENGH > 13 NUCLEO-TIDES
Tm = 4°C x (number of G's and C's in the primer)+2°C x (number of A's and T's in the primer) For salt adjusted Tm calculation,	T_m = 64.9°C + 41°C x (number of G's and C's in the primer – 16.4)/N Where, N is the number of nucleotides in the primer
Tm = T_m = 81.5°C + 16.6°C x (log_{10}[SALT]+0.41°C x (%GC) – 657/N	

3. Primer annealing temperature (T_a): Too high Ta will produce insufficient primer-template hybridization resulting in low PCR product yield while too low T_a will lead to non-specific PCR products caused by a high number of base pair mismatches. Since T_a is the function of T_m so can be calculated with respect to melting temperature as given below:

T_a =0.3 x T_m (primer) + 0.7 T_m (product) – 14.9 Where,
T_m(primer) = Melting Temperature of the primers,
T_m (product) = Melting temperature of the product

4. GC Content: The number of G's and C's in the primer as a percentage of the total bases should be 40-60%.

5. GC clamp: As GC forms a stronger bond than AT, the number of GC content at the 3' end of the primer should not be more than 3 otherwise it will result in non-specific tight binding at regions where G and C are abundant.

6. Primer secondary structures: Primer secondary structures arise as a result of intra or intermolecular attraction within the primer or with other primers which eventually reduce the yield of amplification as the availability of single stranded primers will be limited for PCR. The various types of primer secondary structures are as follows:

Hairpins: Hairpins are loop structures formed by intramolecular interaction within the primer. Optimally a 3' end hairpin with a ΔG of -2 kcal/mol and an internal hairpin with a ΔG of -3 kcal/mol is tolerated generally.

Dimers: A primer dimer is a structure forming a double-strand like structures which is formed by intermolecular interactions between the two primers. If interaction is formed between two homologous or same sense primers, it is called self-dimer whereas if interaction is formed between two different primers, it is called cross-dimer. Optimally, a 3' end self dimer with a ΔG of -5 kcal/mol and an internal self dimer with a ΔG of -6 kcal/mol is tolerated.

Repeats and runs: Repeats are consecutive occurance of di-nucleotide whereas runs are continuous stretch of single nucleotide. A maximum number of repeats and runs accepted is 4 di-nucleotide and 4 base pairs respectively.

Primer-template homology: Primers should be designed in such a way that there should be no homology within the template other than the target site. This will result in non specific binding and amplification.

Analysis of PCR results: Once PCR cycle is complete, the amplified product is loaded in the agarose gel and observed after ethidium bromide staining under UV light sourc. A water blank reaction is included to monitor the cross contaminating DNA source as template. The percentage of agarose gel depends on the size of DNA to be visualized. Generally 0.8-1% agarose gel is used for analyzing 0.5-5 kb amplified DNA while a DNA of larger size or genomic DNA is visualized in gel as low as 0.5%.

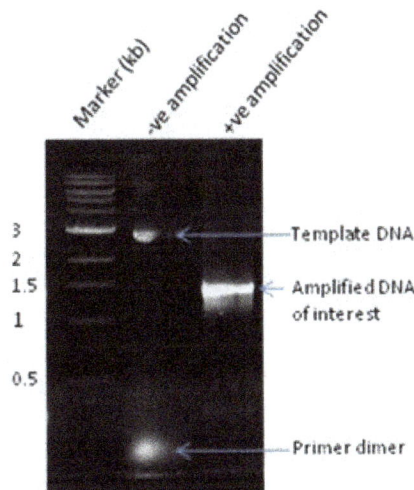

Analysis of PCR product on a agarose gel.

Applications of PCR

PCR in human medicine: PCR technology has become an essential research and diagnostic tool for improving human health and quality of life. It allows the detection of infectious organisms just from one cell by amplifying specific region of the genetic material. Some important areas in medical research where PCR technology is employed include the following:

PCR in infectious disease : PCR technology has become the basis for a broad spectrum of clinical diagnostic tests for various infectious agents, including viruses and bacteria.

Besides detecting the presence of pathogens, PCR also allows us to quantify the amount of pathogens present in patient's blood and this helps to monitor the progression of infection or response to drug treatment. PCR has enabled the development of diagnostic tests for many diseases such as, HIV-1, Hepatitis B and C viruses, Human Papillomavirus, *Chlamydia trachomatis* , *Neisseria gonorrhoeae* , *Cytomegalovirus* and *Mycobacterium tuberculosis* .

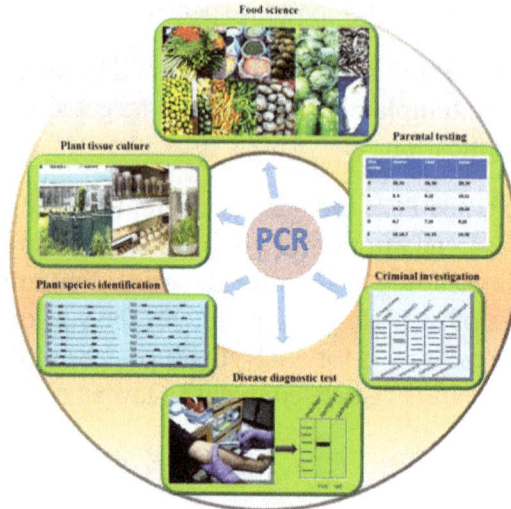

Applications of PCR in various fields of life.

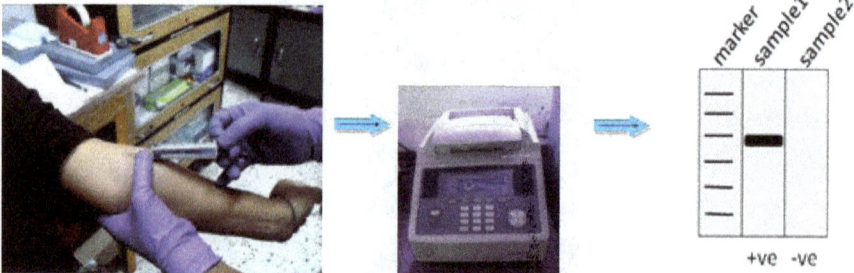

Use of PCR in HIV-1 test. The blood from patient is drawn and the viral DNA is amplied using PCR. The result is shown in gel. Amplification of specific fragment length indicates +ve result while no amplification means −ve result or absence of virus.

PCR and genetic testing: PCR technology has recently become a powerful tool to detect disease associated gene to predict the presence of heart disease and cancers. Knowledge of disease associated gene in the person predisposed to these disorders have a chance to control the problem much in advance.

PCR in plant science: There are various fields in plant science which requires the use of PCR technology for its accomplishment.

Plant species identification: PCR technique has also been employed in identification of plant species using species and group-specific primers targeting chloroplast DNA.

These assays allowed identification of plants based on size-specific amplicons. Plants belonging to the same family has close primer-binding sites and hence same amplicon sizes while plants belonging to different species and groups have different primer-binding sites and hence will result in different amplicon sizes.

PCR in tissue culture: It is used in analysis of DNA and specific genes in plant cells at different stages of re-generation during in vitro culture along with RAPD (random amplification of polymorphic DNA) technology. The level of polymorphism in regenerated plants could be revealed by these dual techniques.

PCR in veterinary parasitology: Owing to its rapidity and sensitivity as compared to antibody-based diagnosis, PCR met its uses in almost all aspects of biological work including veterinary clinical diagnosis. Some examples of PCR applications in veterinary parasitology :-

Aujeszky's disease (pseudorabis) virus of pigs: This virus causes abortion and mortality in piglets. This disease has a latent period where there is no symptom of infection making it difficult to eradicate the disease completely. For this reason, PCR is considered to be appropriate tool for detecting latent cases of Aujeszky's disease.

Bovine leukemia virus (BLV): This virus causes enzootic bovine leukosis. PCR assay for detection of BLV was developed in 1991.

Bovine viral diarrhoea virus (BVDV): This virus is not only fatal to cattle but also causes contamination in calf serum used in cell culture work thus leading to contamination of vaccines and pharmaceutical products. Besides the above examples, PCR has been used in routine diagnosis of veterinary virus such as Porcine parvo, Bovine papilloma type 1 and 2, Avian polyoma, Chicken anemia, Duck hepatitis, African swine fever, Channel catfish, Equine herpes type 1 and 4, Feline herpes, Alcelaphine herpes type1, etc.

PCR in Forensic applications: The most common use of PCR in forensic applications includes:

Use of PCR in criminal investigation. The DNA of suspects is amplified and the digested products are analyzed. The DNA from the crime scene matches that of suspect 3.

Criminal investigation: A sample of blood, hair root or tissue left in the crime scene can be used to identify a person using PCR by comparing the DNA of the crime scene with that of suspect or with DNA database of earlier convicts. Evidence from decades-old crimes can be tested, confirming or defending the people originally convicted.

Parental testing: PCR technology is also used in finding the biological parents of adopted or kidnapped child where the DNA of a child is matched with close relatives. The actual biological father of a newborn can also be ruled out. In parental testing, short tandem repeats (STR) are used as markers where each person's DNA copies contain two copies of these markers one each from father and mother. These markers differ in length and sometimes sequence.

DNA marker	Mother	Child	Father
A	26,31	26,30	29,30
B	8,9	9,10	10,11
C	14,15	14,16	15,16
D	6,7	7,10	9,10
E	14,16.7	14,15	15,18

Use of PCR in Parental testing is done by comparing the DNA markers (given as A→E for convenience) of mother, child and father. The child shares half of the DNA markers from each of the parents.

PCR in research applications: Biological research requires molecular biology techniques as its starting material and so forth, which cannot be accomplished without the use of PCR.

DNA cloning: PCR helps to amplify specific DNA from a genome and the amplified DNA can be inserted into a vector for transformation and expression. These inserts can further be confirmed by PCR method.

DNA sequencing: PCR assists the task of DNA sequencing from patients with genetic disease mutation.

Sequence-tagged sites: This is a process where PCR is used as an indicator if a particular segment of a gene or genome is present in a particular clone. This application is vital in mapping the cosmid clones to be sequenced by the Human Genome Project.

Phylogeny analysis: The phylogeny of organisms like plants, animals and other lower organisms can be traced out by DNA analysis. The origin of unknown samples like the recovered bones of early men can also be ruled out.

Cloning Vector

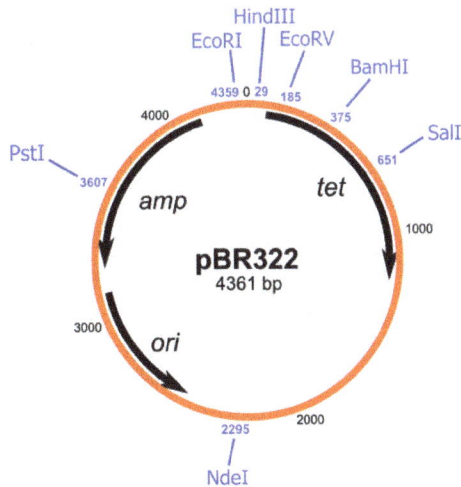

Schematic representation of the pBR322 plasmid, one of the first plasmids widely used as a cloning vector.

A cloning vector is a small piece of DNA, taken from a virus, a plasmid, or the cell of a higher organism, that can be stably maintained in an organism, and into which a foreign DNA fragment can be inserted for cloning purposes. The vector therefore contains features that allow for the convenient insertion or removal of DNA fragment in or out of the vector, for example by treating the vector and the foreign DNA with a restriction enzyme that cuts the DNA. DNA fragments thus generated contain either blunt ends or overhangs known as sticky ends, and vector DNA and foreign DNA with compatible ends can then be joined together by ligation. After a DNA fragment has been cloned into a cloning vector, it may be further subcloned into another vector designed for more specific use.

There are many types of cloning vectors, but the most commonly used ones are genetically engineered plasmids. Cloning is generally first performed using *Escherichia coli*, and cloning vectors in *E. coli* include plasmids, bacteriophages (such as phage λ), cosmids, and bacterial artificial chromosomes (BACs). Some DNA, however, cannot be stably maintained in *E. coli*, for example very large DNA fragments, and other organisms such as yeast may be used. Cloning vectors in yeast include yeast artificial chromosomes (YACs).

Features of a Cloning Vector

All commonly used cloning vectors in molecular biology have key features necessary for their function, such as a suitable cloning site and selectable marker. Others may have additional features specific to their use. For reason of ease and convenience, cloning is often performed using *E. coli*. Thus, the cloning vectors used often have elements necessary for their propagation and maintenance in *E. coli*, such as a functional origin of replication (ori). The ColE1 origin of replication is found in many plasmids. Some vectors also include elements that allow them to be maintained in another organism in addition to *E. coli*, and these vectors are called shuttle vector.

Cloning Site

All cloning vectors have features that allow a gene to be conveniently inserted into the vector or removed from it. This may be a multiple cloning site (MCS) or polylinker, which contains many unique restriction sites. The restriction sites in the MCS are first cleaved by restriction enzymes, then a PCR-amplified target gene also digested with the same enzymes is ligated into the vectors using DNA ligase. The target DNA sequence can be inserted into the vector in a specific direction if so desired. The restriction sites may be further used for sub-cloning into another vector if necessary.

Other cloning vectors may use topoisomerase instead of ligase and cloning may be done more rapidly without the need for restriction digest of the vector or insert. In this TOPO cloning method a linearized vector is activated by attaching topoisomerase I to its ends, and this "TOPO-activated" vector may then accept a PCR product by ligating both the 5' ends of the PCR product, releasing the topoisomerase and forming a circular vector in the process. Another method of cloning without the use of DNA digest and ligase is by DNA recombination, for example as used in the Gateway cloning system. The gene, once cloned into the cloning vector (called entry clone in this method), may be conveniently introduced into a variety of expression vectors by recombination.

Selectable Marker

A selectable marker is carried by the vector to allow the selection of positively transformed cells. Antibiotic resistance is often used as marker, an example being the beta-lactamase gene, which confers resistance to the penicillin group of beta-lactam antibiotics like ampicillin. Some vectors contain two selectable markers, for example the plasmid pACYC177 has both ampicillin and kanamycin resistance gene. Shuttle vector which is designed to be maintained in two different organisms may also require two selectable markers, although some selectable markers such as resistance to zeocin and hygromycin B are effective in different cell types. Auxotrophic selection markers that allow an auxotrophic organism to grow in minimal growth medium may also be used; examples of these are *LEU2* and *URA3* which are used with their corresponding auxotrophic strains of yeast.

Another kind of selectable marker allows for the positive selection of plasmid with cloned gene. This may involve the use of a gene lethal to the host cells, such as barnase, Ccda, and the parD/parE toxins. This typically works by disrupting or removing the lethal gene during the cloning process, and unsuccessful clones where the lethal gene still remains intact would kill the host cells, therefore only successful clones are selected.

Reporter Gene

Reporter genes are used in some cloning vectors to facilitate the screening of successful clones by using features of these genes that allow successful clone to be easily identified. Such features present in cloning vectors may be the *lacZα* fragment for α complementation in blue-white selection, and/or marker gene or reporter genes in frame with and flanking the MCS to facilitate the production of fusion proteins. Examples of fusion partners that may be used for screening are the green fluorescent protein (GFP) and luciferase.

Elements for Expression

A cloning vector need not contain suitable elements for the expression of a cloned target gene, such as a promoter and ribosomal binding site (RBS), many however do, and may then work as an expression vector. The target DNA may be inserted into a site that is under the control of a particular promoter necessary for the expression of the target gene in the chosen host. Where the promoter is present, the expression of the gene is preferably tightly controlled and inducible so that proteins are only produced when required. Some commonly used promoters are the T7 and *lac* promoters. The presence of a promoter is necessary when screening techniques such as blue-white selection are used.

Cloning vectors without promoter and RBS for the cloned DNA sequence are sometimes used, for example when cloning genes whose products are toxic to *E. coli* cells. Promoter and RBS for the cloned DNA sequence are also unnecessary when first making a genomic or cDNA library of clones since the cloned genes are normally subcloned into a more appropriate expression vector if their expression is required.

Some vectors are designed for transcription only with no heterologous protein expressed, for example for *in vitro* mRNA production. These vectors are called transcription vectors. They may lack the sequences necessary for polyadenylation and termination, therefore may not be used for protein production.

Types of Cloning Vectors

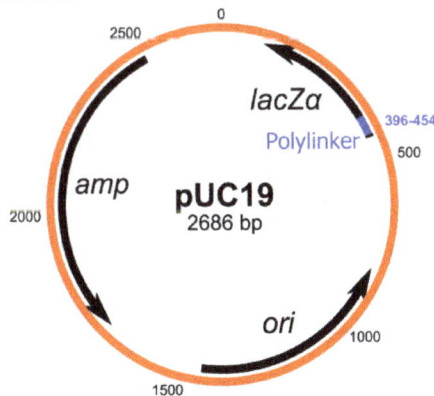

The pUC plasmid has a high copy number, contains a multiple cloning site (polylinker), a gene for ampicillin antibiotic selection, and can be used for blue-white screen.

A large number of cloning vectors are available, and choosing the vector may depend a number of factors, such as the size of the insert, copy number and cloning method. Large insert may not be stably maintained in a general cloning vector, especially for those with a high copy number, therefore cloning large fragments may require more specialized cloning vector.

Plasmid

Plasmids are autonomously replicating circular extra-chromosomal DNA. They are the standard cloning vectors and the ones most commonly used. Most general plasmids may be used to clone DNA insert of up to 15 kb in size. One of the earliest commonly used cloning vectors is the pBR322 plasmid. Other cloning vectors include the pUC series of plasmids, and a large number of different cloning plasmid vectors are available. Many plasmids have high copy number, for example pUC19 which has a copy number of 500-700 copies per cell, and high copy number is useful as it produces greater yield of recombinant plasmid for subsequent manipulation. However low-copy-number plasmids may be preferably used in certain circumstances, for example, when the protein from the cloned gene is toxic to the cells.

Some plasmids contain an M13 bacteriophage origin of replication and may be used to generate single-stranded DNA. These are called phagemid, and examples are the pBluescript series of cloning vectors.

Bacteriophage

The bacteriophages used for cloning are the phage λ and M13 phage. There is an upper limit on the amount of DNA that can be packed into a phage (a maximum of 53 kb), therefore to allow foreign DNA to be inserted into phage DNA, phage cloning vectors may need to have some non-essential genes deleted, for example the genes for lysogeny since using phage λ as a cloning vector involves only the lytic cycle. There are two kinds of λ phage vectors - insertion vector and replacement vector. Insertion vectors contain a unique cleavage site whereby foreign DNA with size of 5–11 kb may be inserted. In replacement vectors, the cleavage sites flank a region containing genes not essential for the lytic cycle, and this region may be deleted and replaced by the DNA insert in the cloning process, and a larger sized DNA of 8–24 kb may be inserted.

There is also a lower size limit for DNA that can be packed into a phage, and vector DNA that is too small cannot be properly packaged into the phage. This property can be used for selection - vector without insert may be too small, therefore only vectors with insert may be selected for propagation.

Cosmid

Cosmids are plasmids that incorporate a segment of bacteriophage λ DNA that has the cohesive end site (cos) which contains elements required for packaging DNA into λ particles. It is normally used to clone large DNA fragments between 28 and 45 Kb.

Bacterial Artificial Chromosome

Insert size of up to 350 kb can be cloned in bacterial artificial chromosome (BAC). BACs are maintained in *E. coli* with a copy number of only 1 per cell. BACs are based on F plasmid, another artificial chromosome called the PAC is based on the P1 phage.

Yeast Artificial Chromosome

Insert of up to 3,000 kb may be carried by yeast artificial chromosome.

Human Artificial Chromosome

Human artificial chromosome may be potentially useful as a gene transfer vectors for gene delivery into human cells, and a tool for expression studies and determining human chromosome function. It can carry very large DNA fragment (there is no upper limit on size for practical purposes), therefore it does not have the problem of limited cloning capacity of other vectors, and it also avoids possible insertional mutagenesis caused by integration into host chromosomes by viral vector.

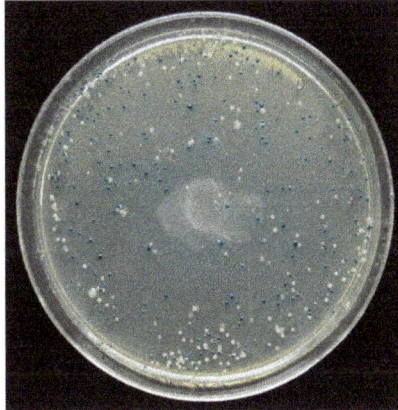

An LB agar plate showing the result of a blue white screen. White colonies may contain an insert in the plasmid it carries, while the blue ones are usuccessful clones.

Screening: Example of the Blue/White Screen

Many general purpose vectors such as pUC19 usually include a system for detecting the presence of a cloned DNA fragment, based on the loss of an easily scored phenotype. The most widely used is the gene coding for *E. coli* β-galactosidase, whose activity can easily be detected by the ability of the enzyme it encodes to hydrolyze the soluble, colourless substrate X-gal (5-bromo-4-chloro-3-indolyl-beta-d-galactoside) into an insoluble, blue product (5,5'-dibromo-4,4'-dichloro indigo). Cloning a fragment of DNA within the vector-based *lacZα* sequence of the β-galactosidase prevents the production of an active enzyme. If X-gal is included in the selective agar plates, transformant colonies are generally blue in the case of a vector with no inserted DNA and white in the case of a vector containing a fragment of cloned DNA.

Eukaryotic Vector

Prokaryotic vectors are good to express the proteins of eukaryotic or prokaryotic origin but in few specific cases, they are not well suited. Such as, eukaryotic protein is unstable, require special environment for folding or losses its the biological activity. In few cases especially in the case of production of therapeutic proteins, cross contamination of bacterial products may cause clinical problems in humans. Eukaryotic vector system is designed to clone and express gene in eukaryotic cells such as yeast (saccharomyces cerevisiae), insect and mammalian cell lines. There are two different types of eukaryotic vectors-

1. Vector as extrachromosomal DNA- These vectors remains in eukaryotic cell as extrachromosomal DNA and express the protein.

2. Integration Vector- These vectors carry an integration site to facilitate recombination medited integration into the chromosomal DNA of the host cell. In general, eukaryotic vector contains origin of replication from bacteria and eukaryotic cells. In addition, they also contain different selectable markers for prokaryotic and eukaryotic cells. These modifications allow to use and perform easy cloning procedure in bacterial host system to generate recombinant construct containing foreign DNA in vector. The basic features required for a vector discussed previously for prokaryotic system is also required for eukaryotic vector as well.

Saccharomyces cerevisiae vector system-There are 3 types of yeast vector system. These all have couple of similar features such as presence of MCS, shuttle vector (origin of replication for *E. coli* and Yeast) and presence of selectable markers.

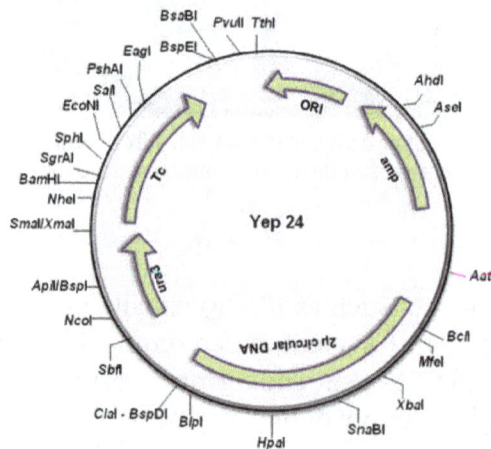

Vector map of episomal yeast plasmid Yep 24.

1. Episomal vector- Yeast episomal vector are constructed by combining bacterial plasmid either with yeast 2μ origin of replication or with autonomous replication sequence (ARS). The yeast vector containing ARS are high copy number plasmid but they are unstable in the absence of selection pressure. This can be over-come

by adding centromeric sequence (CEN) but it affects the plasmid copy number and as a result it become a low copy number plasmid. ARS based yeast plasmids are not very popular for expression of protein. In contrast, 2μ based vector are very popular for heterologous protein expression. A representative 2μ based episomal yeast vector is shown in the figure above. It is a 6.3kb plasmid with a copy number in the range of 50-100 per cell. These plasmids are much more stable than ARS based plasmids.

2. Integrating vector- Episomal yeast vectors are present as extra-chromosomal DNA and are unstable. This can over-come by integration of vector into the host chromosome. In yeast, integration occurs by homologous recombination. The yeast integration plasmids contain target sequence for integration into chromosomal DNA, a selection marker and bacterial origin of replication. Before vector delivery to the yeast, it is digested with the unique restriction endonuclease to produce linear DNA to increase the transformation efficiency and integration. In most of the cases integration is done in such a way that yeast chromosomal DNA remained intact and integration may not affect yeast growth. But in an alternate approach, a portion of yeast chromosomal DNA is replaced with the vector DNA through homologous recombination. These vectors are known as 'transplantment integration vector' and they have foreign DNA, selection marker and homologous DNA to the region of chromosomal DNA to be replaced.

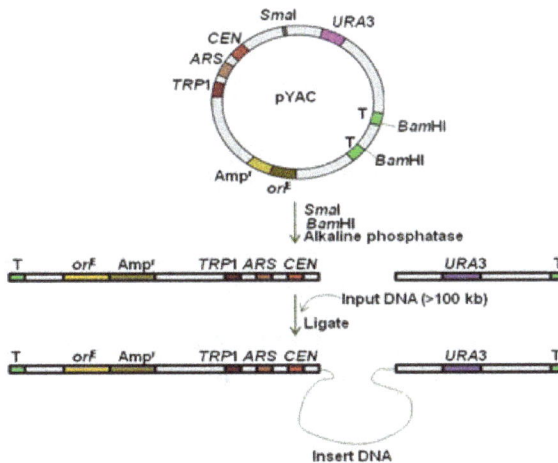

Vector map of YAC plasmid (pYAC) and YAC cloning system.

3. Yeast artificial chromosome- Yeast artificial chromosome (YAC) is the vector of choice used to clone very large DNA fragment (~100kb) to prepare genomic library. YAC vector is like a chromosome as it has ARS sequences, centromere sequence and telomere at the two ends to give stability. A typical YAC plasmid (pYAC) is shown in the figure above. It has an ampicillin resistance gene (Amp^r) for selection in *E. coli* and an *E. coli* origin of replication for propagation in bacteria. In addition, it has ARS for replication, CEN for centromere function, and URA3, TRP1 for

selection in yeast. URA3 and TRP1 is the crucial gene of uracil biosynthesis and tryptophan biosynthesis pathway. For cloning, YAC is digested with SmaI/BamHI, alkaline phosphatase to generate a linear plasmid DNA, now foreign DNA is added for ligation. The recombinant DNA will allow a yeast (Ura-/Trp-) to grow on uracil and tryptophan deficient media.

Baculovirus Vector-Baculovirus is a rod shape virus infecting invertebrate including insect cells. Post infection, virus is either released as free virions or many virus particles are trapped in a protein complex known as polyhedron. The protein responsible for trapping virus into polyhedron is polyhydrin and it help in transmission of virus from one host to other. The polyhydrin is not important for virus propagation but it is under very strong promoter to produce the protein in large quantities. Realizing this fact, replacement of polyhydrin gene with a foreign DNA fragment will allow expression of protein in large quantities. The baculovirus *Autographa californica* multiple nuclear polyhedrosis virus (AcMNPV) is used as a vector to express protein. The transfer vector map of AcMNPV is given in the figure below. The gene of interest will be inserted into the cloning site placed adjacent to the promoter. It has polyhedron termination sequence down-stream to the cloning site to stop transcription of cloned gene.

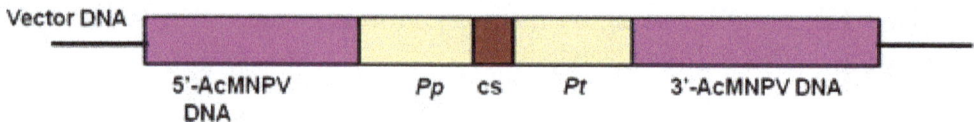

Structural elements of a baculovirus transfer vector.

Mammalian Vector- large number of excellent mammalian vectors are in circulation to clone eukaryotic gene for protein synthesis and study the transcription mechanism. A generalized scheme with the structural elements required to design mammalian vector is given in the figure below. As discussed earlier, it contains a eukaryotic replication of origin from an animal virus such as SV40 from simian virus 40. A promoter to drive the expression of foreign gene and selection marker, other eukaryotic features such as polyadenylation, transcription termination site etc.

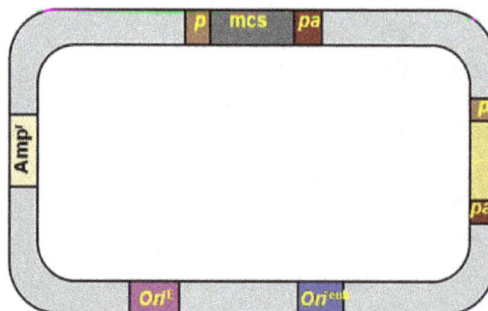

Structural elements of a mammalian expression vector

References

- Von Fange, T.; McDiarmid, T.; MacKler, L.; Zolotor, A. (2008). "Clinical inquiries: Can recombinant growth hormone effectively treat idiopathic short stature?". The Journal of family practice. 57 (9): 611–612. PMID 18786336

- Campbell, Neil A. & Reece, Jane B.. (2002). Biology (6th ed.). San Francisco: Addison Wesley. pp. 375–401. ISBN 0-201-75054-6

- linkAmerican Association for the Advancement of Science (1903). Science. Moses King. pp. 502–. Retrieved 8 October 2010

- Fernandez, M.; Hosey, R. (2009). "Performance-enhancing drugs snare nonathletes, too". The Journal of family practice. 58 (1): 16–23. PMID 19141266

- Watson, James D. (2007). Recombinant DNA: Genes and Genomes: A Short Course. San Francisco: W.H. Freeman. ISBN 0-7167-2866-4

- Yong, Ed (2013-03-15). "Resurrecting the Extinct Frog with a Stomach for a Womb". National Geographic. Retrieved 2013-03-15

- Russell, David W.; Sambrook, Joseph (2001). Molecular cloning: a laboratory manual. Cold Spring Harbor, N.Y: Cold Spring Harbor Laboratory. ISBN 0-87969-576-5

- "Tasmanian bush could be oldest living organism". Discovery Channel. Archived from the original on 23 July 2006. Retrieved 2008-05-07

- Lobban, P.; Kaiser, A. (1973). "Enzymatic end-to end joining of DNA molecules". Journal of Molecular Biology. 78 (3): 453–471. doi:10.1016/0022-2836(73)90468-3. PMID 4754844

- Brown, Terry (2006). Gene Cloning and DNA Analysis: an Introduction. Cambridge, MA: Blackwell Pub. ISBN 1-4051-1121-6

- "Generations of Cloned Mice With Normal Lifespans Created: 25th Generation and Counting". Science Daily. 7 March 2013. Retrieved 8 March 2013

- William Sims Bainbridge, Ph.D. Religious Opposition to Cloning Journal of Evolution and Technology - Vol. 13 - October 2003

- Peter J. Russel (2005). iGenetics: A Molecular Approach. San Francisco, California, United States of America: Pearson Education. ISBN 0-8053-4665-1

- Halim, N. (September 2002). "Extensive new study shows abnormalities in cloned animals". Massachusetts institute of technology. Retrieved October 31, 2011

- Bertani G, Weigle JJ (Feb 1953). "Host controlled variation in bacterial viruses". Journal of Bacteriology. 65 (2): 113–21. PMC 169650. PMID 13034700

- Rantala, Milgram, M., Arthur (1999). Cloning: For and Against. Chicago, Illinois: Carus Publishing Company. p. 1. ISBN 0-8126-9375-2

- Gray, Richard; Dobson, Roger (31 January 2009). "Extinct ibex is resurrected by cloning". The Telegraph. London. Retrieved 2009-02-01

- de Grey, Aubrey; Rae, Michael (September 2007). Ending Aging: The Rejuvenation Breakthroughs that Could Reverse Human Aging in Our Lifetime. New York, NY: St. Martin's Press, 416 pp. ISBN 0-312-36706-6

- Luria SE, Human ML (Oct 1952). "A nonhereditary, host-induced variation of bacterial viruses". Journal of Bacteriology. 64 (4): 557–69. PMC 169391. PMID 12999684

4

Biotechnology: Methods and Techniques

Spectroscopic methods are used to determine protein concentration and estimate DNA and its melting temperature. This section discusses spectroscopic methods and immunological methods as a few procedures under biotechnology. Science and technology have undergone rapid developments in the past decade which has resulted in the discovery of significant tools and techniques in the field of biotechnology; which have been extensively detailed in this chapter.

Spectroscopy

Estimation of protein concentration in a given protein preparation is one of the most commonly performed tasks in a biochemistry lab. There are several ways of estimating the protein concentration such as amino acid analysis following acid hydrolysis of the protein; analyzing the changes in the spectral properties of certain dyes in the presence of proteins; and spectrophotometric estimation of the proteins in near or far UV region. Although dye-binding assays and amino acid analysis following acid hydrolysis of the protein can be used for estimating the protein concentration for both pure as well as an unknown mixture of proteins; UV spectroscopic quantitation holds good for the pure proteins. If a protein is pure, UV spectroscopic quantitation is the method of choice because it is easy and less time-consuming to perform; furthermore, the protein sample can be recovered back.

Absorption of ultraviolet radiation is a general method used for estimating a large number of bioanalytes. The region of the electromagnetic radiation ranging from ~10 – 400 nm is identified as the ultraviolet region. For the sake of convenience in referring to the different energies of UV region, it can be divided into three regions:

- Near UV region (UV region nearest to the visible region; $\lambda \sim 250 - 400$ nm)

- Far UV region (UV region farther to the visible region; $\lambda \sim 190 - 250$ nm)

- Vacuum UV region ($\lambda < 190$ nm)

This division is not strict and you may find slightly different wavelength ranges for these regions. Absorption of UV light is associated with the electronic transitions in the molecules from lower to higher energy states.

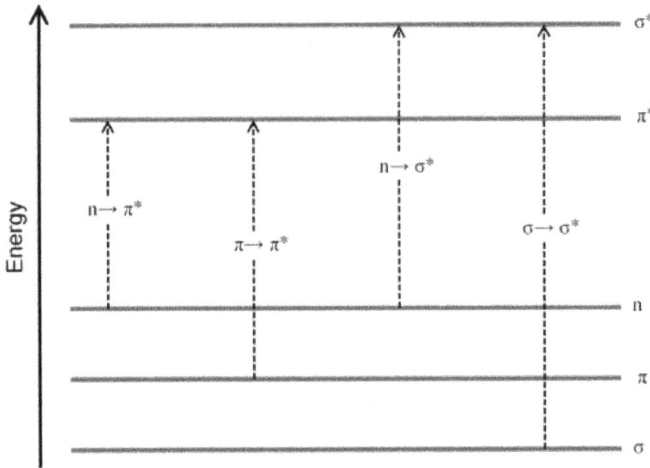

A diagrammatic representation of the energy levels of molecular orbitals; the vertical arrows represent electronic transitions.

As is clear from the figure, $\sigma \to \sigma^*$ transition involves very high energy and usually lies in the vacuum UV region. Saturated hydrocarbons, that can undergo only $\sigma \to \sigma^*$ transition, therefore show absorption bands at ~150 nm wavelength. Compounds that have unsaturation and/or lone pair of electrons *i.e.* the ones that can undergo $\pi \to \pi^*$ or $n \to \pi^*$ transitions, absorb at higher wavelengths that may lie in far or near UV regions, the regions of UV radiation the biochemical spectroscopists are usually interested in. The group of atoms in a molecule that comprise the orbitals involved in the transition is said to constitute a chromophore. The figure below shows an absorption spectrum of a peptide. The spectrum immediately suggests that the proteins can absorb both in near UV and far UV regions.

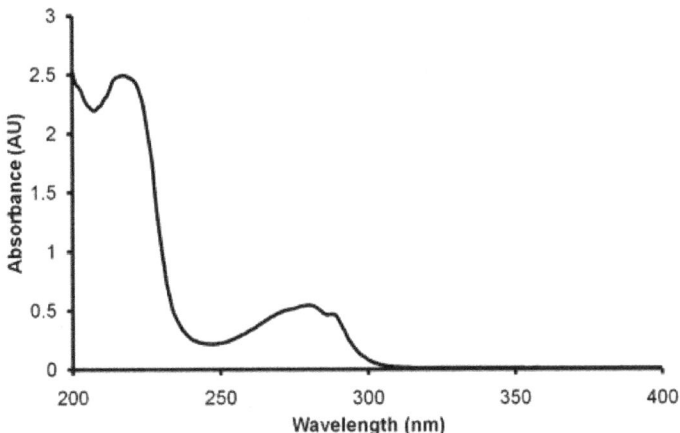

Absorption spectrum of a peptide in the near and far UV regions.

Absorption of UV radiation is usually represented in terms of absorbance and %transmittance:

$$\text{Absorbance (A)} = -\log\left(\frac{I}{I_0}\right)$$

$$\%\text{transmit} \tan ce\ (\%T) = \frac{I}{I_0} \times 100$$

where, I_0 and I represent the intensities of light entering and exiting the sample, respectively.

Absorbance of an analyte depends on the concentration of the analyte and the path length of the solution (Beer-Lambert Law):

$$A = \varepsilon cl$$

where, ε is the molar absorption coefficient, c is the molar concentration of the analyte and l is the path length of the cell containing the analyte solution. If molar absorption coefficient of the analyte and the path length of sample cell are known, concentration can directly be determined using Beer-Lambert law.

Let us see how protein concentration is estimated using near and far UV radiation.

Near-UV Radiation

Aromatic amino acids, tryptophan, tyrosine, and phenylalanine and the disulfide linkage constitute the chromophores that absorb in the near UV region. Absorption of near UV radiation by proteins is usually monitored at 280 nm due to very high absorption by Trp and Tyr at this wavelength. The table below shows the molar absorption coefficient of the protein chromophores that absorb the light of 280 nm.

Table: Molar absorption coefficients of protein chromophores at 280 nm

	$\varepsilon_{280}\ (M^{-1}cm^{-1})$		
	Trp	Tyr	S-S
Average value in folded proteins	5500	1490	125
Value in unfolded proteins	5690	1280	120

where, ε_{280} is the molar absorption coefficient at 280 nm.

It is therefore straightforward to calculate the molar absorption coefficient of a folded protein if its amino acid sequence or composition is known:

$$\varepsilon_{280} = (5500 \times n_{Trp}) + (1490 \times n_{Tyr}) + (125 \times n_{s-s})$$

For short peptides that are usually unfolded in water, the molar absorption coefficients can be calculated using the following equation:

$$\varepsilon_{280} = (5690 \times n_{Trp}) + (1280 \times n_{Tyr}) + (120 \times n_{s-s})$$

Far-UV Radiation

The proteins and peptides that lack aromatic residues and disulfide linkage do not absorb the near UV radiation. The concentration of such proteins and peptides can be estimated using far UV radiation. Peptide bond is the major chromophore in the far UV region with a strong absorption band around 190 nm ($\pi \to \pi^*$ transition) and a weak band around 220 nm ($n \to \pi^*$ transition). As oxygen strongly absorbs 190 nm radiation, it is convenient to measure absorption at 205 nm where molar absorption coefficient of peptide bond is roughly half of that at 190 nm. A 1 mg/ml solution of most proteins would have an extinction coefficient of ~30 – 35 at 205 nm. This means that the result obtained can have more than 15% error. An empirical formula, proposed by Scopes provides the $A_{205}^{1mg/ml}$ within ± 2%:

$$A_{205}^{1mg/ml} = 27 + 120 \left(\frac{A_{280}}{A_{205}} \right)$$

Alternatively, the concentration can be estimated using Wadell's method that relies on the absorbance at 215 and 225 nm:

$$Pr otein\ concentration \left(\frac{ug}{ml} \right) = 144(A_{215} - A_{225})$$

Materials

- A UV/Visible spectrophotometer

- Pipettes

- Pipette tips

- Disposable microfuge tubes

- Quartz cuvettes (suitable for wavelengths smaller than 205 nm)

- Pure protein solution in a buffer (or in water)

- The buffer the protein is dissolved in (will act as the blank).

Procedure

1. Switch 'ON' the UV/visible spectrophotometer and allow it 30 minutes warm up.

2. Determine the number of tryptophans, tyrosines, and disulfide linkages present in the protein.

3. Determine the molar absorption coefficient of the protein at 280 nm using 4.4.

4. Take the buffer used for protein dissolution in the quartz cuvettes.

    a. The volume of buffer has to be sufficient enough to cover the entire aperture the light beam passes through and depends on the capacity of the quartz cuvette; typically cuvettes with 1 *ml* capacity are used.

5. Place the cuvettes in the reference cell and sample cell slots in the spectrophotometer.

6. 'ZERO' the baseline for the 250 – 350 nm range.

7. Remove the quartz cuvette placed in the sample cell slot and discard all the contents.

8. Add the same volume of the given protein solution into the cuvette and place it back in the sample cell slot.

9. Record the absorbance at 280 nm (A_{280}^{sample}) and 330 nm (A_{330}^{sample})

    a. Proteins do not absorb at wavelengths higher than 320 nm; any absorbance obtained at 330 nm therefore arises due to scattering.

    b. If the absorbance at 280 nm does not lie between 0.05 – 1.0, dilute the protein solution in the same buffer so as to obtain an absorbance in this range.

10. Switch off the spectrophotometer.

11. Take out the quartz cell and clean them using detergent solution and deionized water.

Calculation

The absorbance at 280 nm is corrected for light scattering:

$$A_{280(corrected)}^{sample} = A_{280}^{sample} - 1.929 \times (A_{330}^{sample})$$

The amount of the given protein is determined using Beer-Lambert law:

$$A_{280(corrected)}^{sample} = \varepsilon c l$$

$$c(M) = \frac{A_{280(corrected)}^{sample}}{\varepsilon(M^{-1}cm^{-1})l(cm)}$$

Aim

To determine the total protein concentration in a given sample

Introduction:

The concentration of proteins can be estimated using various methods. For estimating the total protein in a complex protein mixture, one can use dyes that exhibit changes in their spectral properties on binding to the proteins. Bradford is one such dye-based assay for protein concentration estimation.

The principle underlying Bradford assay is the binding of the Coomassie Blue G250 dye to proteins.

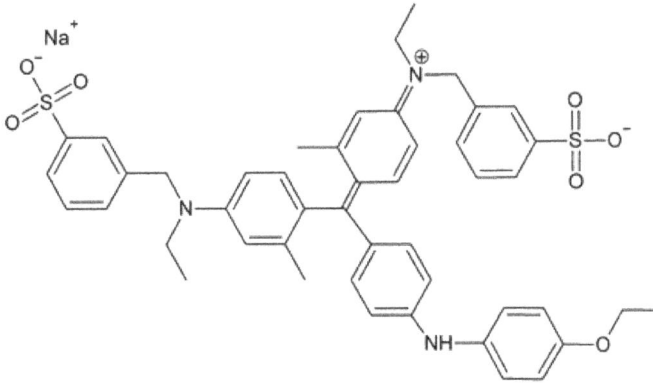

Structure of Coomassie Blue G250

Free Coomassie Blue G250 can exist in four different ionization states with pK_{a1}, pK_{a2}, and pK_{a3} of 1.15, 1.82, and 12.4. At pH 0, both the sulfate groups are negatively charged and all three nitrogens are positively charged giving the dye +1 net charge (the red form of the dye). Around pH 1.5, the neutral green form of the dye predominates. At neutral pH, the dye has a net charge of +1 (the blue form of the dye). The red, green, and blue forms of the dye absorb visible radiation with absorption maxima at 470, 650, and 590 nm, respectively. It is the anionic form of the dye that binds to the protein. Binding of the blue form of Coomassie Blue G250 with proteins causes red-shift in its absorption spectrum; the absorption maximum shifts from 590 to 620 nm. It, therefore, looks sensible to record the absorption at 620 nm. The absorbance, however, is recorded at 595 nm to avoid any contribution from the green form of the dye. The dye binds more readily to the cationic residues, lysine and arginine. This implies that the response of the assay would depend on the amino acid composition of protein, the major drawback of the assay. The original assay developed by Bradford shows such variation between different proteins. Several modifications have been introduced into the assay to overcome this drawback; the modified assays, however, are more susceptible to interference by other chemicals than the original assay. The original Bradford assay, therefore, remains the most convenient and widely used method.

In this experiment, we shall be using the standard Bradford assay which is suitable for measuring the protein amount ranging from 10 – 10 μg. A microassay suitable for the protein ranging from 1–10 μg is also briefly discussed.

Materials

Equipments:

1. A visible range spectrophotometer

2. Vortex mixer

3. Weighing balance

Reagents:

1. Bradford reagent

2. Ovalbumin (Protein standard)

Glassware and plasticware:

1. Pipettes

2. Pipette tips

3. A 5 ml glass pipette

4. Pipette aid

5. 100 ml measuring cylinder

6. Test tubes (for standard assay)

7. 1.5 ml microfuge tubes (for microassay)

8. Plastic cuvettes

Preparation of Reagents:

Bradford reagent : Bradford reagent is prepared as follows:

1. Weigh 200 *mg* of Coomassie Blue G250 dye and dissolve it in 50 *ml* of 95% ethanol.

2. Mix this solution with 100 *ml* of concentrated (85%) phosphoric acid.

3. Make the final volume of the solution to 1 *litre* by adding distilled water.

4. Filter the reagent through Whatman No. 1 filter paper.

5. Transfer the filtrate in an amber colored bottle and store at room temperature.

Protein Standard : Ovalbumin; the Standard Solution is Prepared as follows

1. Weigh accurately 5 *mg* ovalbumin.

2. Dissolve it in 5 *ml* distilled water; this gives a protein stock solution of 1 *mg/ml* concentration.

3. Store the protein standard at −20°C.

Procedure of Standard Bradford Assay

1. Take out the frozen protein standard and allow it to come to room temperature.

2. As the concentration of the unknown protein sample can be anything, the assay will be performed with a range of dilutions (1, 1:10, 1:100, and 1:1000). Prepare 100 *µl* of each of the dilutions.

3. Take 15 test tubes and label them from 1 to 15.

4. Pipette out 10 *µl* , 20 *µl* , 30 *µl* , ..., 100 *µl* ovalbumin standard in the glass tubes labeled 1 – 10; leave blank the tube no. 11.

5. Add distilled water to make the final volume 100 *µl* in each of the tubes (including blank).

6. Take 100 *µl* of each of the unknown protein dilutions in the tubes labeled 12 – 15.

7. Add 5 *ml* of Bradford reagent in each of the tubes and mix well by inversion or gentle vortex mixing (avoid frothing).

8. Within 5 – 60 *min* , measure the absorbance of tubes 1 – 10 and 12 – 15 at 595 nm in the quartz/glass cuvette against the reagent blank (tube 11).

9. Record the readings in the suggested observation table below:

Observation Table

Table: Observation table for the Bradford assay

Tube No.	Volume (μl)	Mass (μg)	Distilled water (μl)	Bradford reagent (ml)	A_{595}
	Standard DNA				
1	10	10	90	5	
2	20	20	80	5	
3	30	30	70	5	
4	40	40	60	5	
5	50	50	50	5	
6	60	60	40	5	
7	70	70	30	5	
8	80	80	20	5	
9	90	90	10	5	
10	100	100	0	5	
11	Blank (0)	0	100	5	
	Unknown sample				
12	100 (1:1000 dil.)	Unknown	0	5	
13	100 (1:100 dil.)	Unknown	0	5	
14	100 (1:10 dil.)	Unknown	0	5	
15	100 (Undiluted)	Unknown	0	5	

Microassay:

1. In the Bradford microassay, the standard protein stock solution is prepared with a concentration of 100 $\mu g/ml$.

2. Make the dilutions of the unknown protein sample exactly as prepared in standard assay.

3. Follow steps 3 – 6 of the standard Bradford assay.

4. Add 1 ml of Bradford reagent in each of the tubes and mix well by inversion or gentle vortex mixing (avoid frothing).

5. Within 5 – 60 min , measure the absorbance of tubes 1 – 10 and 12 – 15 at 595 nm in the quartz/glass cuvette against the reagent blank (tube 11).

6. Record the readings in the observation table.

Analysis: Let us take some hypothetical readings for carrying out the analysis:

Tube No.	Volume (μl)	Mass (μg)	Distilled water (μl)	Bradford reagent (ml)	A_{595}
	Standard DNA				
1	10	10	90	5	0.04
2	20	20	80	5	0.078
3	30	30	70	5	0.122
4	40	40	60	5	0.164
5	50	50	50	5	0.202
6	60	60	40	5	0.240
7	70	70	30	5	0.278
8	80	80	20	5	0.322
9	90	90	10	5	0.361
10	100	100	0	5	0.403
11	Blank (0)	0	100	5	Reference (0)
	Unknown sample				
12	100 (1:1000 dil.)	Unknown	0	5	0.004
13	100 (1:100 dil.)	Unknown	0	5	0.041
14	100 (1:10 dil.)	Unknown	0	5	0.426
15	100 (Undiluted)	Unknown	0	5	2.768

The absorbance values of the standard protein are plotted against the amount of the protein added for the assay. The curve is fit using linear regression with intercept (0,0) as shown in the figure below.

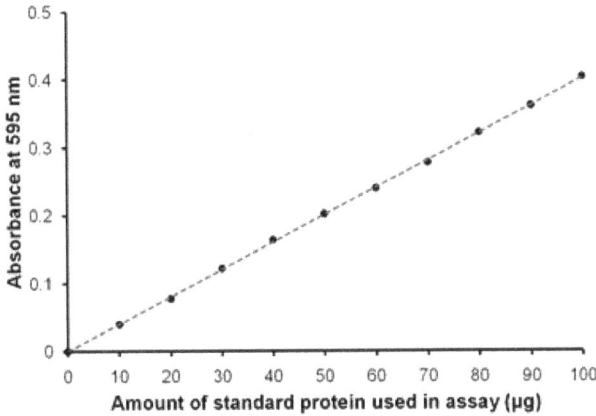

Plot of absorbance at 595 nm against the amount of protein

The equation of this regression line is: *Absorbance = 0.00402 × amount of protein*

$$where\ \textbf{0.00402}\ is\ the\ slope \left(\frac{_\Delta Absorbance}{_\Delta Protein\ amount(ug)} \right)$$

Now, let us calculate the concentration of the unknown protein. We have got absorbance at different dilutions but which one should be used for determining the concentration. The absorbance values lying between 0.05 – 0.6 are most reliable. We shall, therefore, calculate the concentration for the 10-fold diluted sample that gave an absorbance of 0.426. The amount of protein is given by:

$$Protein(ug) = \frac{1}{slope} \times Absorbance$$

$$= \frac{1}{0.00402ug^{-1}} \times 0.426 = 105.97ug$$

As this amount of the protein was present in the 100 μl of the 10-fold diluted protein sample, the concentration of the given protein sample

$$= \frac{105.97\,ug}{100\,ul} \times 10 = 10.597\,ug\,/\,ul \approx 10.6\,mg\,/\,ml$$

The concentration of the unknown sample can directly be calculated as follows:

$$Protein\ concentration = \frac{Absorbance \times dilution\ factor}{slope(\frac{1}{ug}) \times volume\ of\ diluted\ protein\ used\ for\ the\ assay(ul)}$$

$$Protein\ concentration(ug\,/\,ul) = \frac{0.426 \times 10}{0.00402 \times 0.1} = 10597$$

Therefore, the concentration of the protein in the given sample = *10.597 µg/µl = 10.6 mg/ml*

Aim

To determine the concentration of a given DNA sample using diphenylamine method

Introduction:

The principle underlying estimation of DNA using diphenylamine is the reaction of diphenylamine with deoxyribose sugar producing blue-coloured complex. The DNA sample is boiled under extremely acidic conditions; this causes depurination of the DNA followed by dehydration of deoxyribose sugar into a highly reactive ω-hydroxylevulinylaldehyde. The reaction is not specific for DNA and is given by 2-deoxypentoses, in general. The ω-hydroxylevulinylaldehyde, under acidic conditions, reacts with diphenylamine to produce a blue-coloured complex that absorbs at 595 nm. The mechanism of reaction of deoxyribose sugar with diphenylamine is shown in the figure below. As the sugar linked to only purine residues participates in the reaction, the readout is only from 50% of the total number of nucleotides. As this holds true for both the known standard and the given unknown sample, the concentration of the unknown sample can be directly calculated from the standard graph.

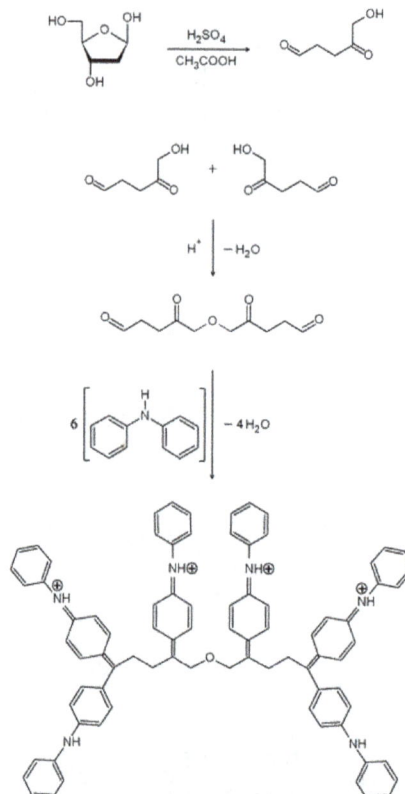

The reaction mechanism of diphenylamine reagent with deoxyribose sugar

Materials

Equipments:

1. A UV/Visible spectrophotometer

2. Vortex mixer

3. Weighing balance

4. Water bath

Reagents:

1. Diphenylamine reagent

2. Calf thymus DNA

3. Glacial acetic acid

4. Concentrated sulfuric acid

Glassware and plasticware:

1. Pipettes

2. Pipette tips

3. A 5 *ml* glass pipette

4. Pipette aid

5. A 100 *ml* measuring cylinder

6. A 250 *ml* amber coloured glass bottle

7. Test tubes

8. Caps for glass tubes

9. Distilled water

10. Quartz or glass cuvettes

Preparation of Reagents

Diphenylamine (DPA) reagent :

1. Weigh 1*g* of diphenylamine and transfer it into a 250 *ml* amber coloured glass bottle.

2. Add 100 *ml* glacial acetic acid and shake well to achieve complete dissolution.

3. Add 2.5 ml of concentrated sulfuric acid.

4. Store the reagent in dark at 2 − 8°C.

Calf thymus DNA (100 μg/ml)

Prepare 100 $μg/ml$ of calf thymus DNA solution in distilled water.

Procedure:

1. As the concentration of the unknown DNA sample can be anything, the assay will be performed with a range of dilutions (1, 1:10, 1:100, and 1:1000). Prepare 1 ml of each of the dilutions.

2. Take 15 test tubes and label them from 1 to 15.

3. Pipette out 100 $μl$, 200 $μl$, 300 $μl$, ..., 1000 $μl$ calf thymus DNA standard in the glass tubes labeled 1 − 10; leave blank the tube no. 11.

4. Add distilled water to make the final volume 1 ml in each of the tubes (including blank).

5. Take 1 ml of each of the unknown DNA dilutions in the tubes labeled 12 − 15.

6. Add 3 ml of DPA reagent in each of the 15 tubes and mix well by vortexing.

7. Cover each of the tubes with the caps and place them in boiling water bath for 10 minutes.

8. Take out all the tubes from water bath and allow them to return to room temperature.

9. Measure the absorbance of tubes 1 − 10 and 12 − 15 at 595 nm in the quartz/glass cuvette against the reagent blank (tube 11).

10. Record the readings in the suggested observation table below:

Observation Table

Table: Observation table for the diphenylamine assay

Tube No.	Volume ($μl$)	Mass ($μg$)	Distilled water ($μl$)	Diphenylamine reagent (ml)	A_{595}
Standard DNA					
1	100	10	900	3	
2	200	20	800	3	
3	300	30	700	3	
4	400	40	600	3	
5	500	50	500	3	
6	600	60	400	3	
7	700	70	300	3	
8	800	80	200	3	
9	900	90	100	3	
10	1000	100	0	3	

11	Blank (0)	0	1000	3	
Unknown sample					
12	1000 (1:1000 dil.)	Unknown	0	3	
13	1000 (1:100 dil.)	Unknown	0	3	
14	1000 (1:10 dil.)	Unknown	0	3	
15	1000 (Undiluted)	Unknown	0	3	

Aim

To determine the melting temperature (T_m) of a given DNA sample using ultraviolet absorption.

A double-helical DNA is made up of two strands that run antiparallel to each other. Each adenine (A) in one strand is paired with a thymine (T) on the other; similarly, each guanine (G) on one strand is paired with a cytosine (C) on the other. A–T and G–C are said to constitute the complementary base pairs. This pairing is achieved through stacking interactions and hydrogen bonding between the bases and is the basis of the double stranded DNA structure and its stability. Heating disrupts these non-covalent interactions between the bases; this could unwind the two strands separating the two strands apart. Separation of the two DNA strands is termed as denaturation or melting of DNA. In the double-helical structure, guanine forms three hydrogen bonds with cytosine while adenine forms two hydrogen bonds with thymine. It is therefore evident that the amount of heat required for denaturing the DNA would depend on its nucleotide composition. The temperature at which 50% of the DNA gets denatured is termed as its melting temperature (T_m).

Nucleic acids absorb very strongly in the near UV region. The absorbance is attributed to the heterocyclic rings present in the nucleotides. At neutral pH, DNA would typically absorb with an absorption maximum around 260 nm. Denaturation of DNA leads to higher absorption of ultraviolet radiation (hyperchromicity). The melting temperature of DNA can therefore be determined simply by monitoring its absorbance at 260 nm while heating it.

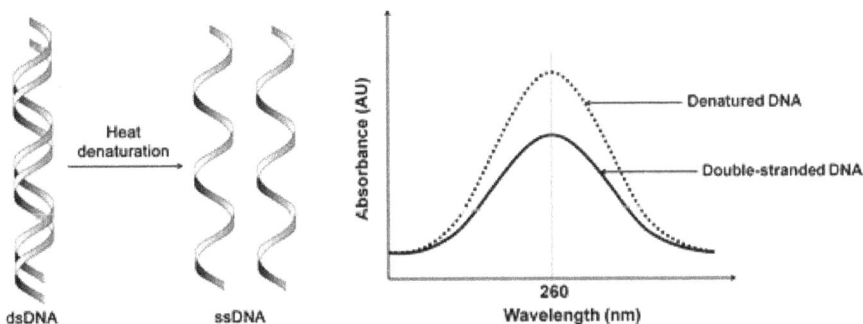

The hyperchromic effect in DNA; denaturation leads to higher absorption

Experimentally, the absorbance of the DNA molecule remains fairly constant at lower temperatures giving a plateau. As the temperature increases, the AT rich regions start melting thereby causing an increase in absorbance. Further increase in temperature causes steep rise in the absorbance followed by another plateau as the DNA gets completely denatured at these temperatures.

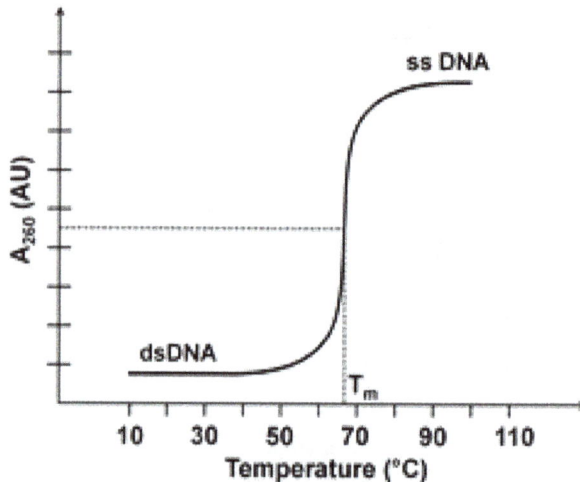

A typical DNA melting curve; the temperature at which half of the DNA is denatured is termed as the melting temperature (T_m).

Materials

Equipments:

1.   A UV/Visible spectrophotometer equipped with Peltier temperature cell:

Peltier accessory is used for achieving very high temperature accuracies. The temperature of the cells or cell holders can be monitored by placing temperature sensors. It is recommended to place the temperature sensors in the cells for achieving more accurate results.

2. Water baths (in case equipment is not equipped with Peltier temperature control)

Glassware and plasticware:

1. Pipettes

2. Pipette tips

3. Parafilm

4. 1.5 *ml* microfuge tubes

5. Distilled water

6. 1 *ml* Quartz cuvettes (Note 1)

7. The buffer the given DNA is dissolved in

Procedure

1. Switch "ON" the spectrophotometer.

2. Set the measurement mode to 'Absorbance' and wavelength to 260 nm.

3. Set the temperature to 20°C in the Peltier attachment.

4. Measure the absorbance of the given DNA sample at 260 nm against the buffer used for dissolving the DNA.

5. An absorbance between 0.1 – 0.4 is suitable for determining the melting tempera-ture. If the absorbance of the DNA is above 0.4, dilute the sample in the given buffer so as to achieve 1 ml DNA solution with an absorbance between 0.2 – 0.3.

6. Measure the absorbance of the diluted sample at 25°C.

7. Increase the temperature by 5°C and measure the absorbance when the cell reaches the specified temperature.

8. Repeat step 7 until a temperature of 95°C is achieved.

9. Record the measurements in the observation table shown below.

Alternative Procedure, in case Peltier Accessory is not there

1. Prepare sufficient volume (at least 15 ml) of the DNA sample in the given buffer so as to obtain an absorbance between 0.1 – 0.4.

2. Take fourteen 1.5 ml microfuge tubes and label them 1 – 14.

3. Add 1 ml of DNA solution into each of the microfuge tubes.

4. Tightly seal all the tubes with parafilm.

5. Label the three water baths as I, II, and III.

6. We shall be measuring absorbance at 20, 30, 40, 45, 50, 55, 60, 65, 70, 75, 80, 85, 90, and 95 degrees Celsius $i.e.$ at 14 different temperatures.

7. Set water baths I, II, and III at 20 °C, 30 °C, and 40 °C temperatures, respectively.

8. Place the tubes 1, 2, and 3 in water baths I, II, and III, respectively and incubate at least for 10 minutes.

9. Take tube 1 out and immediately measure its absorbance at 260 nm against the buffer blank.

10. Set the water bath I to 45 °C and place tube 4 in it once the specified temperature is achieved.

11. Meanwhile, take out tube 2 and measure its absorbance.

12. Set the water bath II to 50 °C and place tube 5 in it once the specified temperature is achieved.

13. Meanwhile, take out tube 3 and measure its absorbance.

14. Set the water bath III to 55 °C and place tube 6 in it once the specified temperature is achieved.

15. This cycle is to be followed until the absorbance is recorded for all the 14 tubes.

16. Record the measurements in the observation table shown below.

Aim

Monitoring equilibrium unfolding of protein using tryptophan fluorescence

Introduction:

Folding of a protein into its unique 3-dimensional structure is central for its function. The tertiary structure of a protein is determined by various intramolecular non-covalent interactions such as H-bonding, electrostatic interactions, and hydrophobic interactions. The conformational stability of the protein structure is an important parameter that defines and limits its utility.

Folding/unfolding of small globular proteins closely approaches the two state folding/unfolding mechanisms:

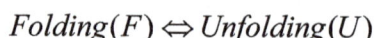

$$Folding(F) \Leftrightarrow Unfolding(U)$$

The conformational stability of a small globular protein can be determined by calculating the equilibrium constant and the free energy, ΔG for the reaction shown in equation (8.1). The value of ΔG for the unfolding reaction shown in the above equation in the absence of a denaturant is referred to as the conformational stability of a protein at a given temperature and is represented by ΔG (H_2O). Comparison of conformational stability of a protein with its variants allows determination of various forces and factors responsible for the protein's stability.

Methods of Unfolding

The native structure of a protein is sensitive to its environment such as pH, temperature, ionic strength, cosolvents, and presence of denaturants. A change in any of these parameters can disrupt the non-covalent interactions thereby causing unfolding (denaturation) of protein. The conformational stability of a protein is most routinely determined by thermal denaturation or by denaturing the protein with urea or guanidinium chloride. Urea solutions have historically been used for determining the conformational stabilities of proteins. Although guanidinium chloride is a stronger denaturant and chemically more stable than urea, it is not preferred over urea because it is a salt and causes changes in the ionic strength of the solutions that could result in less reliable ΔG (H_2O).

Methods for Following Unfolding

Unfolding of a protein can be studied by a variety of methods. The techniques that are more routinely used include ultraviolet difference spectroscopy, fluorescence spectroscopy, and circular dichroism spectroscopy. Unfolding can also be monitored using NMR spectroscopy, measuring biological activity of the protein, viscosity, and optical rotatory dispersion. Fluorescence spectroscopy and circular dichroism spectroscopy are perhaps the two most commonly used methods for monitoring protein unfolding and we shall be discussing the unfolding experiment keeping these two techniques in mind. To decide upon the technique to be used, the spectra of both folded and unfolded proteins need to be recorded. Following spectral features are then considered for deciding upon the technique to be used:

1. The magnitude of the response: At a given concentration of a protein, fluorescence intensity is usually much larger than the ellipticity. The sample amount may therefore be criteria for determining the method for monitoring unfolding. Furthermore, if the given protein lacks tyrosine and tryptophan residues, the fluorescence spectroscopy can simply not be employed.

2. The difference in response for folded and unfolded protein: The fluorescence intensity of the folded protein inthe figure below, for example, is ~4-fold more than that of the unfolded protein at ~322 nm. In general, the wavelength where maximum difference is observed is used. The difference in magnitude may be largest at ~195 nm in far-UV circular dichroism spectra, it is however convenient to monitor unfolding at 220 nm as oxygen absorbs very strongly below 200 nm.

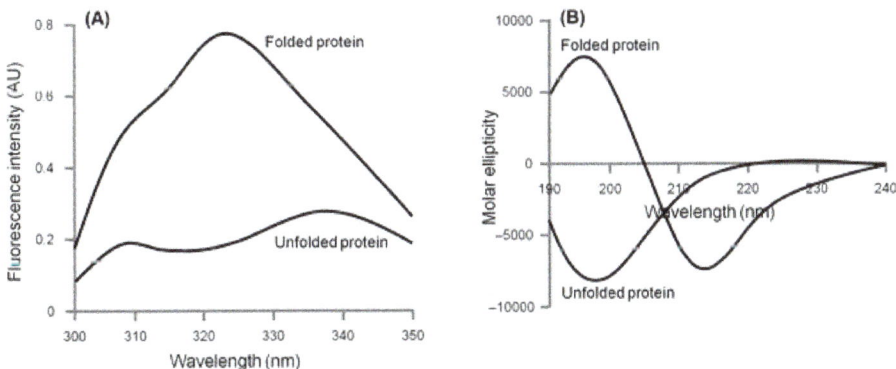

Fluorescence and circular dichroism spectra of a hypothetical protein in folded and unfolded state

3. Signal to noise ratio: Apart from the difference in magnitude in the response, signal

to noise ratio is an additional factor in determining the wavelength.

4. Finally, fluorescence spectroscopy is not recommended for monitoring thermal denaturation as pre- and post-transition baselines are steep and sensitive to temperature.

Materials

Equipments:

1. Spectrofluorometer

2. Weighing balance

3. pH meter

Reagents:

1. Urea

2. 3-(N-Morpholino)propanesulfonic acid sodium salt (MOPS sodium salt)

3. 1 *M* Hydrochloric acid

4. Given protein (RNase T1, commercially available)

Glassware and plasticware:

1. Pipettes

2. Pipette tips

3. 100 *ml* volumetric flasks

4. 100 *ml* beaker

5. Test tubes or 15 *ml* polypropylene tubes

6. Quartz cuvettes

Preparation of Reagents

Urea stock solution : Urea stock solution is prepared as follows:

1. Take a 100 *ml* volumetric flask, place it on the weighing balance, allow the reading to stabilize, and then tare it.

2. Weigh accurately 60 *g* of urea and 0.694 *g* of MOPS sodium salt and add them to a 100 *ml* beaker. For the sake of doing calculations is step 7, let us assume that the weight of the urea was 59.95 *g*.

3. Add 1.8 ml of 1 M HCl and 45 ml of distilled water and allow the urea and MOPS salt to dissolve.

4. Measure the pH of the solution; if required, adjust the pH using 1 M HCl (note down the added mass).

5. Transfer the contents of the beaker into the 'tared' 100 ml volumetric flask.

6. Add distilled water to make the final volume to 100 ml and weigh the volumetric flask. Let us assume that the weight is 115.07 g.

7. Calculate the urea concentration as follows:

   a. Calculate the ratio, $\dfrac{weight\ of\ urea}{weight\ of\ the\ solution}$, let us call this ratio, W.

   Here, $W = \frac{59.95}{115.07} = 0.521$

   b. If d is the density of the solution and d_o is the density of water, then

   $$\frac{d}{d_0} = 1 + 0.2658W + 0.0330W^2$$
   $$= 1 + 0.2658 \times 0.521 + 0.0330 \times (0.521)^2$$
   $$= 1.147$$

   c. The volume of the solution,

   $$V = \frac{weight\ of\ the\ solution}{d/d_0} = \frac{115.07}{1.147} = 100.32 ml$$

   d. Therefore, the molarity of urea

   $$= \frac{weight\ of\ the\ urea}{Molecular\ weight\ of\ the\ urea} \times \frac{1000}{V\,(ml)}$$
   $$= \frac{59.95}{60.056} \times \frac{1000}{100.32} = 9.95M$$

8. This gives a 9.95 M urea solution in 30 mM MOPS buffer, pH 7.0.

MOPS buffer :

1. Weigh 0.694 g of MOPS sodium salt and transfer to a 100 ml beaker.

2. Add 90 ml of distilled water and allow the salt to dissolve completely.

3. Adjust the pH to 7.0 using 1 N HCl.

4. Transfer the contents to a 100 *ml* volumetric flask and add distilled water to make the final volume 100 *ml* .

Protein stock solution :

1. Weigh accurately 100 *mg* of RNase T1 in a 15 *ml* polypropylene tube.

2. Add 10 *ml* of 30 *mM* MOPS buffer, pH 7.0.

3. This gives a 10 *mg/ml* solution of RNase T1 in 30 *mM* MOPS buffer, pH 7.0.

4. Store the protein stock solution at −20 °C.

Procedure

1. Take out the frozen RNase T1 stock solution and allow it to come to room temperature.

2. Take 25 test tubes and label them from 1 to 25.

3. Add increasing volumes of urea stock solution, decreasing volumes of MOPS buffer and a fixed volume of protein stock solution as shown inthe table.

4. Allow the solutions to equilibrate.

5. Switch ON the spectrofluorometer and allow it 30 *min* warm up.

6. Set the excitation wavelength to 280 nm and emission wavelength to 320 nm (*Note 2*).

7. Measure the fluorescence emission for each of the samples at 90° for 30 seconds (this gives multiple readings, depending on the integration time used for each reading).

8. Calculate the average fluorescence reading for each of the samples from the multiple readings obtained in step 7 and record them in the observation table.

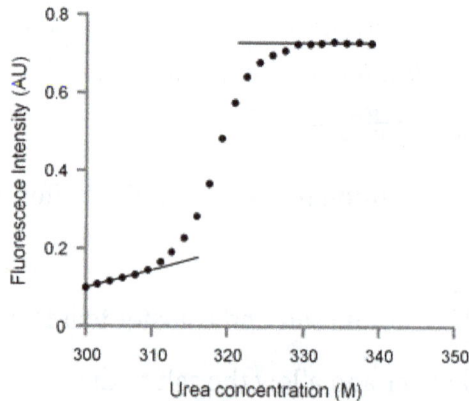

A plot between fluorescence intensity against urea concentration showing a typical two-state protein unfolding curve

9. Plot the fluorescence emission intensity against urea concentration as shown in the figure.

Observation Table

Table: Observation table for the protein unfolding using fluorescence spectroscopy

Tube No.	MOPS buffer (ml)	Urea stock solution (ml)	Protein stock solution (ml)	Urea concentration (M)	Fluorescence intensity at 320 nm (AU)
1	2.8	0	0.2	0	-
2	2.7	0.1	0.2	0.33	-
3	2.6	0.2	0.2	0.67	-
4	2.5	0.3	0.2	1	-
⠿	⠿	⠿	⠿	⠿	-
⠿	⠿	⠿	⠿	⠿	-
⠿	⠿	⠿	⠿	⠿	-
25	0.4	2.4	0.2	8	-

Analysis of Unfolding Curve

For a two-state unfolding process, plotting the fluorescence intensity gives a curve as shown in the figure. A least square fitting is performed on the data to obtain a continuous curve.

For a two-state folding/unfolding mechanism, only folded and unfolded protein states are populated at significant concentrations at any of the urea concentration. Therefore,

$$f_F + f_U = 1$$

where, f_F and f_U represent the fractions of the folded and unfolded proteins, respectively.

Thus the observed value of y (fluorescence intensity, in this case) at any point in the graph is given by:

$$y = y_F f_F + y_U f_U$$

where, y_F and y_U represent the values of y characteristic of the folded and unfolded protein states, respectively and can be calculated from the unfolding curve.

Combining the above equations:

$$f_U = \frac{(y_F - y)}{(y_F - y_U)}$$

The unfolding curve can be divided into three regions:

i. *Pre-transition region* : it shows how y for the folded protein *i.e.* y_F responds to the denaturant.

ii. *Transition region* : it shows how y varies as the unfolding takes place.

iii. *Post-transition region* : it shows how y for the unfolded protein *i.e.* y_U responds to the denaturant.

The equilibrium constant, K_{eq} for the reaction can be calculated as follows:

$$K_{eq} = \frac{f_U}{(1-f_U)} = \frac{f_U}{f_F} = \frac{(y_F - y)}{(y - y_U)}$$

The free energy change, $\Delta\Delta G$ for the reaction can be calculated from the equation:

$$\Delta G = -RT \ln K = -RT \ln \left[\frac{(y_F - y)}{(y - y_U)} \right]$$

The values of K_{eq} are most accurately measured near the midpoint of the denaturation curve and the errors become substantial for values outside the range 0.1 – 10. This corresponds to the ΔG values between −5.7 to +5.7 $kJ/mole$ (~ ±1.36 $kcal/mol$). The data for this range can be tabulated as shown in the table.

Table: Analysis of the urea denaturation curve

Urea concentration (M)	y	f_U	K_{eq}	ΔG
			>0.1	
			<10.0	

In the limited region where ΔG is most accurately measured, it varies linearly with the concentration of denaturant.

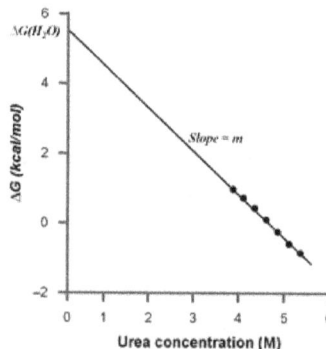

ΔG as a function of denaturant (urea) concentration. Intercept at ΔG gives the stability of the protein, ΔG (H_2O).

The equation of line is obtained from the least square analysis. ΔG in the absence of denaturant *i.e.* ΔG (H_2O) is calculated by extrapolating the line to zero denaturant concentration. The linear equation therefore is given by:

$$\Delta G = \Delta G(H_2O) - m[denaturant]$$

Notes:

1. As we are interested in determining the thermodynamic parameters in the unfolding reaction, it is important to ensure that the unfolding reactions have reached the equilibria before measurements are made. The equilibration time varies from protein to protein and depends on the temperature at which the reaction is being carried out; it can lie anywhere between seconds to days. Equilibrium for RNase T1, for instance, is achieved in minutes at 30 °C but takes hours at 20 °C. The solutions in the pre- and post-transitions regions equilibrate faster than those in the transition region. For an unknown protein, it is necessary to carry out a pilot study for determining the urea concentration corresponding to the transition region and the equilibration times through a pilot study.

2. The optimum emission wavelength may vary from protein to protein and should be determined. For RNase T1, 320 nm is the optimum wavelength.

Circular Dichroism

Circular dichroism (CD) is dichroism involving circularly polarized light, i.e., the differential absorption of left- and right-handed light. Left-hand circular (LHC) and right-hand circular (RHC) polarized light represent two possible spin angular momentum states for a photon, and so circular dichroism is also referred to as dichroism for spin angular momentum. This phenomenon was discovered by Jean-Baptiste Biot, Augustin Fresnel, and Aimé Cotton in the first half of the 19th century. It is exhibited in the absorption bands of optically active chiral molecules. CD spectroscopy has a wide range of applications in many different fields. Most notably, UV CD is used to investigate the secondary structure of proteins. UV/Vis CD is used to investigate charge-transfer transitions. Near-infrared CD is used to investigate geometric and electronic structure by probing metal d→d transitions. Vibrational circular dichroism, which uses light from the infrared energy region, is used for structural studies of small organic molecules, and most recently proteins and DNA.

Physical Principles

Circular Polarization of Light

Electromagnetic radiation consists of an electric (E) and magnetic (B) field that oscillate perpendicular to one another and to the propagating direction, a transverse wave.

While linearly polarized light occurs when the electric field vector oscillates only in one plane, circularly polarized light occurs when the direction of the electric field vector rotates about its propagation direction while the vector retains constant magnitude. At a single point in space, the circularly polarized-vector will trace out a circle over one period of the wave frequency, hence the name. The two diagrams below show the electric vectors of linearly and circularly polarized light, at one moment of time, for a range of positions; the plot of the circularly polarized electric vector forms a helix along the direction of propagation (k). For left circularly polarized light (LCP) with propagation towards the observer, the electric vector rotates counterclockwise. For right circularly polarized light (RCP), the electric vector rotates clockwise.

Linearly polarized

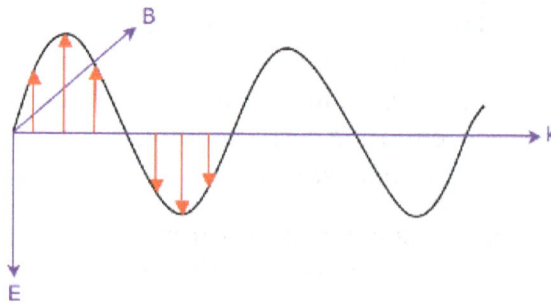

Circularly polarized

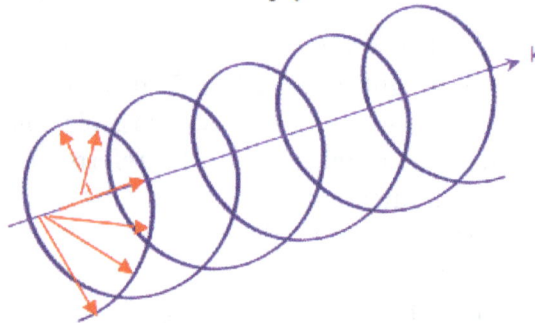

Interaction of Circularly Polarized Light with Matter

When circularly polarized light passes through an absorbing optically active medium, the speeds between right and left polarizations differ ($c_L \neq c_R$) as well as their wavelength ($\lambda_L \neq \lambda_R$) and the extent to which they are absorbed ($\varepsilon_L \neq \varepsilon_R$). *Circular dichroism* is the difference $\Delta\varepsilon \equiv \varepsilon_L - \varepsilon_R$. The electric field of a light beam causes a linear displacement of charge when interacting with a molecule (electric dipole), whereas its magnetic field causes a circulation of charge (magnetic dipole). These two motions combined cause an excitation of an electron in a helical motion, which includes translation and rotation and their associated operators. The experimentally determined relationship between the rotational strength (R) of a sample and the $\Delta\varepsilon$ is given by

$$R_{exp} = \frac{3hc10^3 \ln(10)}{32\pi^3 N_A} \int \frac{\Delta\epsilon}{\nu} d\nu$$

The rotational strength has also been determined theoretically,

$$R_{theo} = \frac{1}{2mc} Im \int \Psi_g \widehat{M}_{(elec.dipole)} \Psi_e d\tau \cdot \int \Psi_g \widehat{M}_{(mag.dipole)} \Psi_e d\tau$$

We see from these two equations that in order to have non-zero $\Delta\epsilon$, the electric and magnetic dipole moment operators ($\widehat{M}_{(elec.dipole)}$ and $\widehat{M}_{(mag.dipole)}$) must transform as the same irreducible representation. C_n and D_n are the only point groups where this can occur, making only chiral molecules CD active.

Simply put, since circularly polarized light itself is "chiral", it interacts differently with chiral molecules. That is, the two types of circularly polarized light are absorbed to different extents. In a CD experiment, equal amounts of left and right circularly polarized light of a selected wavelength are alternately radiated into a (chiral) sample. One of the two polarizations is absorbed more than the other one, and this wavelength-dependent difference of absorption is measured, yielding the CD spectrum of the sample. Due to the interaction with the molecule, the electric field vector of the light traces out an elliptical path after passing through the sample.

It is important that the chirality of the molecule can be conformational rather than structural. That is, for instance, a protein molecule with a helical secondary structure can have a CD that changes with changes in the conformation.

Delta Absorbance

By definition,

$$\Delta A = A_L - A_R$$

where ΔA (Delta Absorbance) is the difference between absorbance of left circularly polarized (LCP) and right circularly polarized (RCP) light (this is what is usually measured). ΔA is a function of wavelength, so for a measurement to be meaningful the wavelength at which it was performed must be known.

Molar Circular Dichroism

It can also be expressed, by applying Beer's law, as:

$$\Delta A = (\epsilon_L - \epsilon_R)Cl$$

where

ε_L and ε_R are the molar extinction coefficients for LCP and RCP light,

C is the molar concentration,

l is the path length in centimeters (cm).

Then

$$\Delta\epsilon = \epsilon_L - \epsilon_R$$

is the molar circular dichroism. This intrinsic property is what is usually meant by the circular dichroism of the substance. Since $\Delta\epsilon$ is a function of wavelength, a molar circular dichroism value ($\Delta\epsilon$) must specify the wavelength at which it is valid.

Extrinsic Effects on Circular Dichroism

In many practical applications of circular dichroism (CD), as discussed below, the measured CD is not simply an intrinsic property of the molecule, but rather depends on the molecular conformation. In such a case the CD may also be a function of temperature, concentration, and the chemical environment, including solvents. In this case the reported CD value must also specify these other relevant factors in order to be meaningful.

Molar Ellipticity

Although ΔA is usually measured, for historical reasons most measurements are reported in degrees of ellipticity. Molar ellipticity is circular dichroism corrected for concentration. Molar circular dichroism and molar ellipticity, [θ], are readily interconverted by the equation:

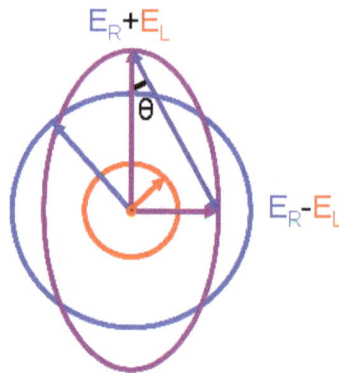

Elliptical polarized light (violet) is composed of unequal contributions of right (blue) and left (red) circular polarized light.

$$[\theta] = 3298.2\Delta\varepsilon.$$

This relationship is derived by defining the ellipticity of the polarization as:

$$\tan\theta = \frac{E_R - E_L}{E_R + E_L}$$

where

E_R and E_L are the magnitudes of the electric field vectors of the right-circularly and left-circularly polarized light, respectively.

When E_R equals E_L (when there is no difference in the absorbance of right- and left-circular polarized light), θ is $0°$ and the light is linearly polarized. When either E_R or E_L is equal to zero (when there is complete absorbance of the circular polarized light in one direction), θ is $45°$ and the light is circularly polarized.

Generally, the circular dichroism effect is small, so $\tan\theta$ is small and can be approximated as θ in radians. Since the intensity or irradiance, I, of light is proportional to the square of the electric-field vector, the ellipticity becomes:

$$\theta(\text{radians}) = \frac{(I_R^{1/2} - I_L^{1/2})}{(I_R^{1/2} + I_L^{1/2})}$$

Then by substituting for I using Beer's law in natural logarithm form:

$$I = I_0 e^{-A\ln 10}$$

The ellipticity can now be written as:

$$\theta(\text{radians}) = \frac{(e^{\frac{-A_R}{2}\ln 10} - e^{\frac{-A_L}{2}\ln 10})}{(e^{\frac{-A_R}{2}\ln 10} + e^{\frac{-A_L}{2}\ln 10})} = \frac{e^{\Delta A \frac{\ln 10}{2}} - 1}{e^{\Delta A \frac{\ln 10}{2}} + 1}$$

Since $\Delta A \ll 1$, this expression can be approximated by expanding the exponentials in a Taylor series to first-order and then discarding terms of ΔA in comparison with unity and converting from radians to degrees:

$$\theta(\text{degrees}) = \Delta A \left(\frac{\ln 10}{4}\right)\left(\frac{180}{\pi}\right)$$

The linear dependence of solute concentration and pathlength is removed by defining molar ellipticity as,

$$[\theta] = \frac{100\theta}{Cl}$$

Then combining the last two expression with Beer's law, molar ellipticity becomes:

$$[\theta] = 100\Delta\varepsilon \left(\frac{\ln 10}{4} \right) \left(\frac{180}{\pi} \right) = 3298.2\Delta\varepsilon$$

The units of molar ellipticity are historically (deg·cm²/dmol). To calculate molar ellipticity, the sample concentration (g/L), cell pathlength (cm), and the molecular weight (g/mol) must be known.

If the sample is a protein, the mean residue weight (average molecular weight of the amino acid residues it contains) is often used in place of the molecular weight, essentially treating the protein as a solution of amino acids. Using mean residue ellipticity facilitates comparing the CD of proteins of different molecular weight; use of this normalized CD is important in studies of protein structure.

Mean Residue Ellipticity

Methods for estimating secondary structure in polymers, proteins and polypeptides in particular, often require that the measured molar ellipticity spectrum be converted to a normalized value, specifically a value independent of the polymer length. Mean residue ellipticity is used for this purpose; it is simply the measured molar ellipticity of the molecule divided by the number of monomer units (residues) in the molecule.

Application to Biological Molecules

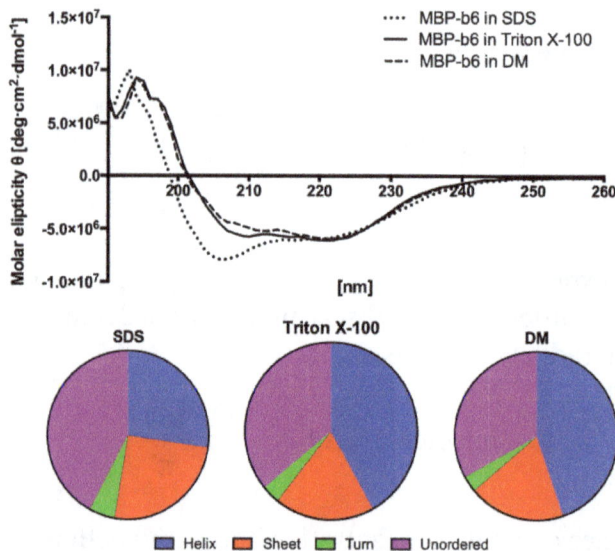

Upper panel: Circular dichroism spectroscopy in the ultraviolet wavelength region (UV-CD) of MBP-cytochrome b_6 fusion protein in different detergent solutions. It shows that the protein in DM, as well as in Triton X-100 solution, recovered its structure. However the spectra obtained from SDS solution shows decreased ellipticity in the range between 200–210 nm, which indicates incomplete secondary structure recovery. Lower panel: The content of secondary structures predicted from the CD spectra using the CDSSTR algorithm. The protein in SDS solution shows increased content of unordered structures and decreased helices content.

In general, this phenomenon will be exhibited in absorption bands of any optically active molecule. As a consequence, circular dichroism is exhibited by biological molecules, because of their dextrorotary and levorotary components. Even more important is that a secondary structure will also impart a distinct CD to its respective molecules. Therefore, the alpha helix of proteins and the double helix of nucleic acids have CD spectral signatures representative of their structures. The capacity of CD to give a representative structural signature makes it a powerful tool in modern biochemistry with applications that can be found in virtually every field of study.

CD is closely related to the optical rotatory dispersion (ORD) technique, and is generally considered to be more advanced. CD is measured in or near the absorption bands of the molecule of interest, while ORD can be measured far from these bands. CD's advantage is apparent in the data analysis. Structural elements are more clearly distinguished since their recorded bands do not overlap extensively at particular wavelengths as they do in ORD. In principle these two spectral measurements can be interconverted through an integral transform (Kramers–Kronig relation), if all the absorptions are included in the measurements.

The far-UV (ultraviolet) CD spectrum of proteins can reveal important characteristics of their secondary structure. CD spectra can be readily used to estimate the fraction of a molecule that is in the alpha-helix conformation, the beta-sheet conformation, the beta-turn conformation, or some other (e.g. random coil) conformation. These fractional assignments place important constraints on the possible secondary conformations that the protein can be in. CD cannot, in general, say where the alpha helices that are detected are located within the molecule or even completely predict how many there are. Despite this, CD is a valuable tool, especially for showing changes in conformation. It can, for instance, be used to study how the secondary structure of a molecule changes as a function of temperature or of the concentration of denaturing agents, e.g. Guanidinium chloride or urea. In this way it can reveal important thermodynamic information about the molecule (such as the enthalpy and Gibbs free energy of denaturation) that cannot otherwise be easily obtained. Anyone attempting to study a protein will find CD a valuable tool for verifying that the protein is in its native conformation before undertaking extensive and/or expensive experiments with it. Also, there are a number of other uses for CD spectroscopy in protein chemistry not related to alpha-helix fraction estimation.

The near-UV CD spectrum (>250 nm) of proteins provides information on the tertiary structure. The signals obtained in the 250–300 nm region are due to the absorption, dipole orientation and the nature of the surrounding environment of the phenylalanine, tyrosine, cysteine (or S-S disulfide bridges) and tryptophan amino acids. Unlike in far-UV CD, the near-UV CD spectrum cannot be assigned to any particular 3D structure. Rather, near-UV CD spectra provide structural information on the nature of the prosthetic groups in proteins, e.g., the heme groups in hemoglobin and cytochrome c.

Visible CD spectroscopy is a very powerful technique to study metal–protein inter-actions and can resolve individual d–d electronic transitions as separate bands. CD spectra in the visible light region are only produced when a metal ion is in a chiral environment, thus, free metal ions in solution are not detected. This has the advantage of only observing the protein-bound metal, so pH dependence and stoichiometries are readily obtained. Optical activity in transition metal ion complexes have been attribut-ed to configurational, conformational and the vicinal effects. Klewpatinond and Viles (2007) have produced a set of empirical rules for predicting the appearance of visible CD spectra for Cu^{2+} and Ni^{2+} square-planar complexes involving histidine and main-chain coordination.

CD gives less specific structural information than X-ray crystallography and protein NMR spectroscopy, for example, which both give atomic resolution data. However, CD spectroscopy is a quick method that does not require large amounts of proteins or ex-tensive data processing. Thus CD can be used to survey a large number of solvent con-ditions, varying temperature, pH, salinity, and the presence of various cofactors.

CD spectroscopy is usually used to study proteins in solution, and thus it complements methods that study the solid state. This is also a limitation, in that many proteins are embedded in membranes in their native state, and solutions containing membrane structures are often strongly scattering. CD is sometimes measured in thin films.

Experimental Limitations

CD has also been studied in carbohydrates, but with limited success due to the experi-mental difficulties associated with measurement of CD spectra in the vacuum ultravio-let (VUV) region of the spectrum (100–200 nm), where the corresponding CD bands of unsubstituted carbohydrates lie. Substituted carbohydrates with bands above the VUV region have been successfully measured.

Measurement of CD is also complicated by the fact that typical aqueous buffer systems often absorb in the range where structural features exhibit differential absorption of circularly polarized light. Phosphate, sulfate, carbonate, and acetate buffers are gener-ally incompatible with CD unless made extremely dilute e.g. in the 10–50 mM range. The TRIS buffer system should be completely avoided when performing far-UV CD. Borate and Onium compounds are often used to establish the appropriate pH range for CD experiments. Some experimenters have substituted fluoride for chloride ion because fluoride absorbs less in the far UV, and some have worked in pure water. An-other, almost universal, technique is to minimize solvent absorption by using shorter path length cells when working in the far UV, 0.1 mm path lengths are not uncommon in this work.

In addition to measuring in aqueous systems, CD, particularly far-UV CD, can be mea-sured in organic solvents e.g. ethanol, methanol, trifluoroethanol (TFE). The latter has the advantage to induce structure formation of proteins, inducing beta-sheets in some and alpha helices in others, which they would not show under normal aqueous conditions. Most common organic solvents such as acetonitrile, THF, chloroform, dichloromethane are however, incompatible with far-UV CD.

It may be of interest to note that the protein CD spectra used in secondary structure estimation are related to the Π to Π* orbital absorptions of the amide bonds linking the amino acids. These absorption bands lie partly in the *so-called* vacuum ultraviolet (wavelengths less than about 200 nm). The wavelength region of interest is actually inaccessible in air because of the strong absorption of light by oxygen at these wavelengths. In practice these spectra are measured not in vacuum but in an oxygen-free instrument (filled with pure nitrogen gas).

Once oxygen has been eliminated, perhaps the second most important technical factor in working below 200 nm is to design the rest of the optical system to have low losses in this region. Critical in this regard is the use of aluminized mirrors whose coatings have been optimized for low loss in this region of the spectrum.

The usual light source in these instruments is a high pressure, short-arc xenon lamp. Ordinary xenon arc lamps are unsuitable for use in the low UV. Instead, specially constructed lamps with envelopes made from high-purity synthetic fused silica must be used.

Light from synchrotron sources has a much higher flux at short wavelengths, and has been used to record CD down to 160 nm. In 2010 the CD spectrophotometer at the electron storage ring facility ISA at the University of Aarhus in Denmark was used to record solid state CD spectra down to 120 nm. At the quantum mechanical level, the feature density of circular dichroism and optical rotation are identical. Optical rotary dispersion and circular dichroism share the same quantum information content.

Infrared Spectroscopy

Infrared spectroscopy (IR spectroscopy or Vibrational Spectroscopy) involves the interaction of infrared radiation with matter. It covers a range of techniques, mostly based on absorption spectroscopy. As with all spectroscopic techniques, it can be used to identify and study chemicals. For a given sample which may be solid, liquid, or gaseous, the method or technique of infrared spectroscopy uses an instrument called an infrared spectrometer (or spectrophotometer) to produce an infrared spectrum. A basic IR spectrum is essentially a graph of infrared light absorbance (or transmittance) on the vertical axis vs. frequency or wavelength on the horizontal axis. Typical units of frequency used in IR spectra are reciprocal centimeters (sometimes called wave numbers), with the symbol cm^{-1}. Units of IR wavelength are commonly given in micrometers (formerly called "microns"), symbol μm, which are related to wave numbers in a reciprocal way. A common laboratory instrument that uses this technique is a Fourier transform infrared (FTIR) spectrometer. Two-dimensional IR is also possible as discussed below.

The infrared portion of the electromagnetic spectrum is usually divided into three regions; the near-, mid- and far- infrared, named for their relation to the visible spectrum. The higher-energy near-IR, approximately 14000–4000 cm^{-1} (0.8–2.5 μm wavelength) can excite overtone or harmonic vibrations. The mid-infrared, approximately 4000–400 cm^{-1} (2.5–25 μm) may be used to study the fundamental vibrations and associated rotational-vibrational structure. The far-infrared, approximately 400–10 cm^{-1} (25–1000 μm), lying adjacent to the microwave region, has low energy and may be used for rotational spectroscopy. The names and classifications of these subregions are conventions, and are only loosely based on the relative molecular or electromagnetic properties.

Theory

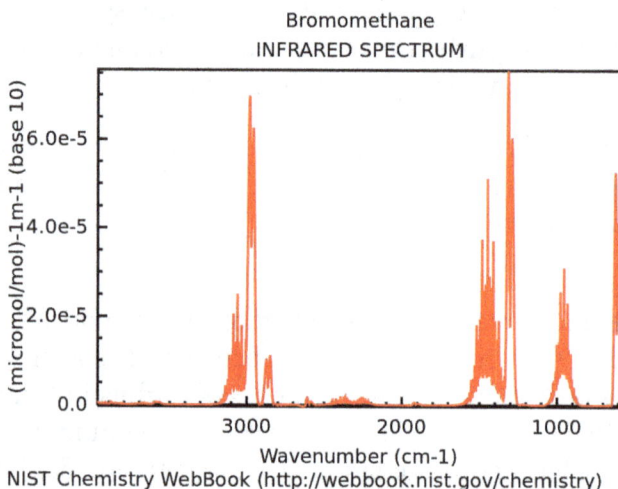

Bromomethane
INFRARED SPECTRUM

NIST Chemistry WebBook (http://webbook.nist.gov/chemistry)

Sample of an IR spec. reading; this one is from bromomethane (CH$_3$Br), showing peaks around 3000, 1300, and 1000 cm^{-1} (on the horizontal axis).

Infrared spectroscopy exploits the fact that molecules absorb frequencies that are characteristic of their structure. These absorptions are resonant frequencies, i.e. the frequency of the absorbed radiation matches the vibrational frequency. The energies are affected by the shape of the molecular potential energy surfaces, the masses of the atoms, and the associated vibronic coupling.

In particular, in the Born–Oppenheimer and harmonic approximations, i.e. when the molecular Hamiltonian corresponding to the electronic ground state can be approximated by a harmonic oscillator in the neighborhood of the equilibrium molecular geometry, the resonant frequencies are associated with the normal modes corresponding to the molecular electronic ground state potential energy surface. The resonant frequencies are also related to the strength of the bond and the mass of the atoms at either end of it. Thus, the frequency of the vibrations are associated with a particular normal mode of motion and a particular bond type.

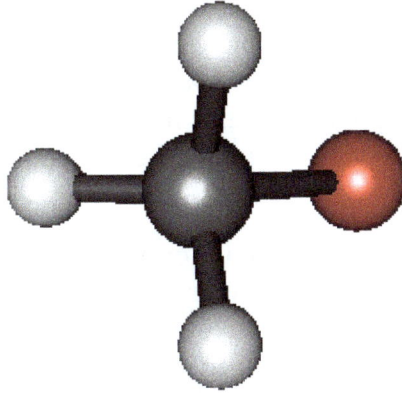

The symmetric stretching of the C–H bonds of bromomethane

Number of Vibrational Modes

In order for a vibrational mode in a sample to be "IR active", it must be associated with changes in the dipole moment. A permanent dipole is not necessary, as the rule requires only a change in dipole moment.

A molecule can vibrate in many ways, and each way is called a *vibrational mode*. For molecules with N number of atoms, linear molecules have $3N - 5$ degrees of vibrational modes, whereas nonlinear molecules have $3N - 6$ degrees of vibrational modes (also called vibrational degrees of freedom). As an example H_2O, a non-linear molecule, will have $3 \times 3 - 6 = 3$ degrees of vibrational freedom, or modes.

TYPES OF VIBRATIONS

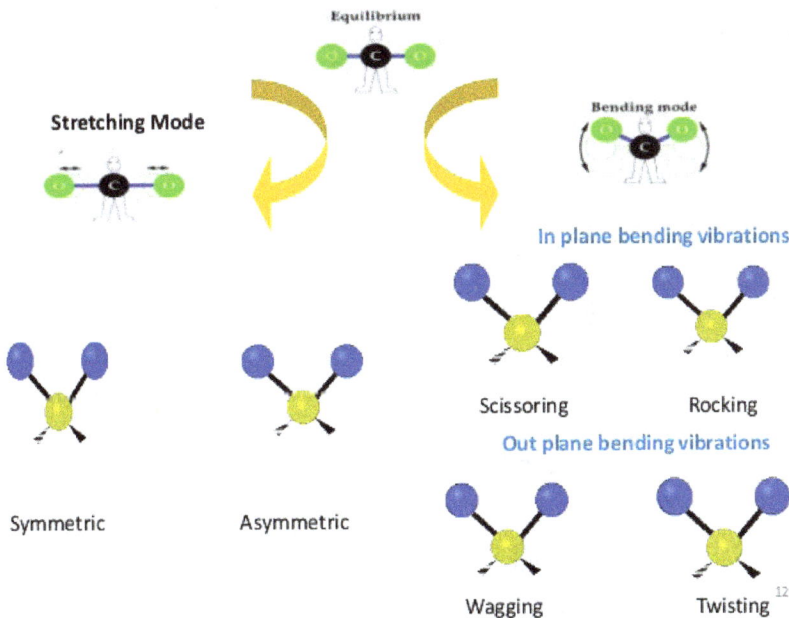

Simple diatomic molecules have only one bond and only one vibrational band. If the molecule is symmetrical, e.g. N_2, the band is not observed in the IR spectrum, but only in the Raman spectrum. Asymmetrical diatomic molecules, e.g. CO, absorb in the IR spectrum. More complex molecules have many bonds, and their vibrational spectra are correspondingly more complex, i.e. big molecules have many peaks in their IR spectra.

The atoms in a CH_2X_2 group, commonly found in organic compounds and where X can represent any other atom, can vibrate in nine different ways. Six of these vibrations involve only the CH_2 portion: symmetric and antisymmetric stretching, scissoring, rocking, wagging and twisting, as shown below. Structures that do not have the two additional X groups attached have fewer modes because some modes are defined by specific relationships to those other attached groups. For example, in water, the rocking, wagging, and twisting modes do not exist because these types of motions of the H represent simple rotation of the whole molecule rather than vibrations within it.

These figures do not represent the "recoil" of the C atoms, which, though necessarily present to balance the overall movements of the molecule, are much smaller than the movements of the lighter H atoms.

Special Effects

The simplest and most important IR bands arise from the "normal modes," the simplest distortions of the molecule. In some cases, "overtone bands" are observed. These bands arise from the absorption of a photon that leads to a doubly excited vibrational state. Such bands appear at approximately twice the energy of the normal mode. Some vibrations, so-called 'combination modes," involve more than one normal mode. The phenomenon of Fermi resonance can arise when two modes are similar in energy; Fermi resonance results in an unexpected shift in energy and intensity of the bands etc.

Practical IR Spectroscopy

The infrared spectrum of a sample is recorded by passing a beam of infrared light through the sample. When the frequency of the IR is the same as the vibrational frequency of a bond or collection of bonds, absorption occurs. Examination of the transmitted light reveals how much energy was absorbed at each frequency (or wavelength). This measurement can be achieved by scanning the wavelength range using a monochromator. Alternatively, the entire wavelength range is measured using a Fourier transform instrument and then a transmittance or absorbance spectrum is generated using a dedicated procedure.

This technique is commonly used for analyzing samples with covalent bonds. Simple spectra are obtained from samples with few IR active bonds and high levels of purity.

More complex molecular structures lead to more absorption bands and more complex spectra.

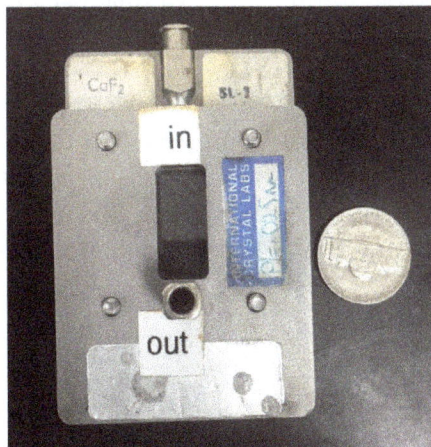

Typical IR solution cell. The windows are CaF_2.

Sample Preparation

Gaseous samples require a sample cell with a long pathlength to compensate for the diluteness. The pathlength of the sample cell depends on the concentration of the compound of interest. A simple glass tube with length of 5 to 10 cm equipped with infra-red-transparent windows at the both ends of the tube can be used for concentrations down to several hundred ppm. Sample gas concentrations well below ppm can be measured with a White's cell in which the infrared light is guided with mirrors to travel through the gas. White's cells are available with optical pathlength starting from 0.5 m up to hundred meters.

Liquid samples can be sandwiched between two plates of a salt (commonly sodium chloride, or common salt, although a number of other salts such as potassium bromide or calcium fluoride are also used). The plates are transparent to the infrared light and do not introduce any lines onto the spectra.

Solid samples can be prepared in a variety of ways. One common method is to crush the sample with an oily mulling agent (usually mineral oil Nujol). A thin film of the mull is applied onto salt plates and measured. The second method is to grind a quantity of the sample with a specially purified salt (usually potassium bromide) finely (to remove scattering effects from large crystals). This powder mixture is then pressed in a mechanical press to form a translucent pellet through which the beam of the spectrometer can pass. A third technique is the "cast film" technique, which is used mainly for polymeric materials. The sample is first dissolved in a suitable, non hygroscopic solvent. A drop of this solution is deposited on surface of KBr or NaCl cell. The solution is then evaporated to dryness and the film formed on the cell is analysed directly. Care is important to ensure that the film is not too thick otherwise light cannot pass through.

This technique is suitable for qualitative analysis. The final method is to use microtomy to cut a thin (20–100 μm) film from a solid sample. This is one of the most important ways of analysing failed plastic products for example because the integrity of the solid is preserved.

In photoacoustic spectroscopy the need for sample treatment is minimal. The sample, liquid or solid, is placed into the sample cup which is inserted into the photoacoustic cell which is then sealed for the measurement. The sample may be one solid piece, powder or basically in any form for the measurement. For example, a piece of rock can be inserted into the sample cup and the spectrum measured from it.

Comparing to a Reference

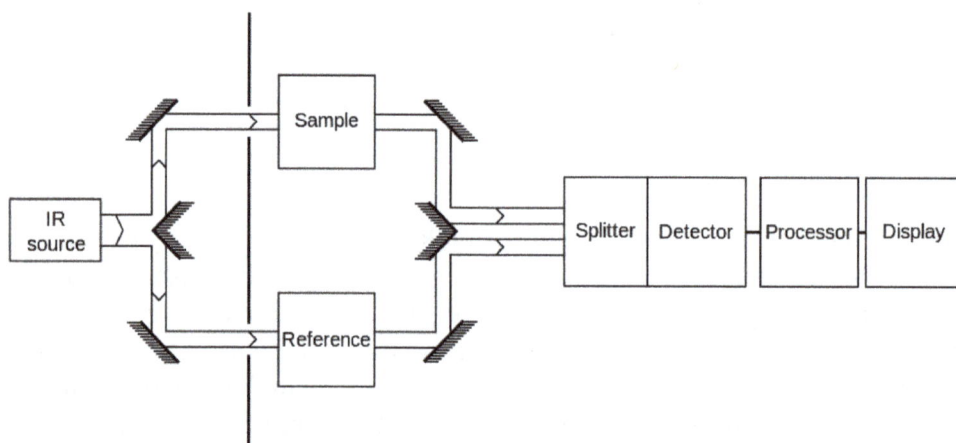

Schematics of a two-beam absorption spectrometer. A beam of infrared light is produced, passed through an interferometer (not shown), and then split into two separate beams. One is passed through the sample, the other passed through a reference. The beams are both reflected back towards a detector, however first they pass through a splitter, which quickly alternates which of the two beams enters the detector. The two signals are then compared and a printout is obtained. This "two-beam" setup gives accurate spectra even if the intensity of the light source drifts over time.

It is typical to record spectrum of both the sample and a "reference". This step controls for a number of variables, e.g. infrared detector, which may affect the spectrum. The reference measurement makes it possible to eliminate the instrument influence.

The appropriate "reference" depends on the measurement and its goal. The simplest reference measurement is to simply remove the sample (replacing it by air). However, sometimes a different reference is more useful. For example, if the sample is a dilute solute dissolved in water in a beaker, then a good reference measurement might be to measure pure water in the same beaker. Then the reference measurement would cancel out not only all the instrumental properties (like what light source is used), but also the light-absorbing and light-reflecting properties of the water and beaker, and the final result would just show the properties of the solute (at least approximately).

A common way to compare to a reference is sequentially: first measure the reference, then replace the reference by the sample and measure the sample. This technique is not perfectly reliable; if the infrared lamp is a bit brighter during the reference measurement, then a bit dimmer during the sample measurement, the measurement will be distorted. More elaborate methods, such as a "two-beam" setup, can correct for these types of effects to give very accurate results. The Standard addition method can be used to statistically cancel these errors.

Nevertheless, among different absorption based techniques which are used for gaseous species detection, Cavity ring-down spectroscopy (CRDS) can be used as a calibration free method. The fact that CRDS is based on the measurements of photon life-times (and not the laser intensity) makes it needless for any calibration and comparison with a reference.

FTIR

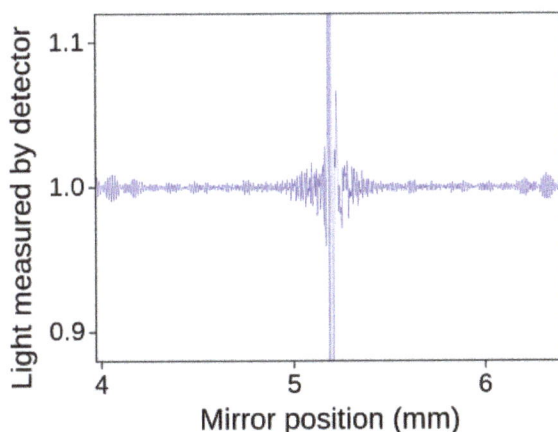

An interferogram from an FTIR measurement. The horizontal axis is the position of the mirror, and the vertical axis is the amount of light detected. This is the "raw data" which can be Fourier transformed to get the actual spectrum.

Fourier transform infrared (FTIR) spectroscopy is a measurement technique that allows one to record infrared spectra. Infrared light is guided through an interferometer and then through the sample (or vice versa). A moving mirror inside the apparatus alters the distribution of infrared light that passes through the interferometer. The signal directly recorded, called an "interferogram", represents light output as a function of mirror position. A data-processing technique called Fourier transform turns this raw data into the desired result (the sample's spectrum): Light output as a function of infrared wavelength (or equivalently, wavenumber). As described above, the sample's spectrum is always compared to a reference.

An alternate method for acquiring spectra is the "dispersive" or "scanning monochromator" method. In this approach, the sample is irradiated sequentially with various single wavelengths. The dispersive method is more common in UV-Vis spectroscopy, but is less practical in the infrared than the FTIR method. One reason that FTIR is favored is called "Fellgett's advantage" or the "multiplex advantage": The information at all frequencies is collected simultaneously, improving both speed and signal-to-noise ratio. Another is called "Jacquinot's Throughput Advantage": A dispersive measurement requires detecting much lower light levels than an FTIR measurement. There are other advantages, as well as some disadvantages, but virtually all modern infrared spectrometers are FTIR instruments.

Absorption Bands

IR spectroscopy is often used to identify structures because functional groups give rise to characteristic bands both in terms of intensity and position (frequency). The positions of these bands are summarized in correlation tables as shown below.

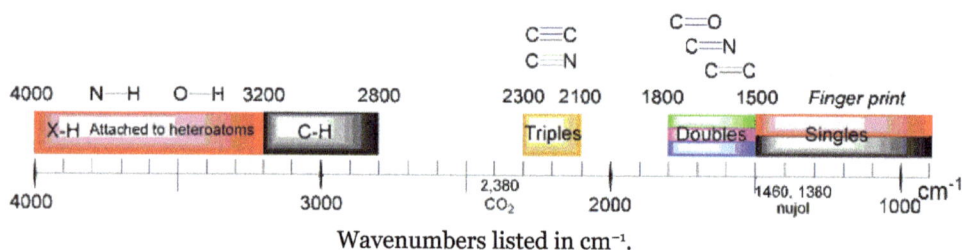

Wavenumbers listed in cm^{-1}.

Badger's Rule

For many kinds of samples, the assignments are known, i.e. which bond deformation(s) are associated with which frequency. In such cases further information can be gleaned about the strength on a bond, relying on the empirical guideline called Badger's Rule. Originally published by Richard Badger in 1934, this rule states that the strength of a bond correlates with the frequency of its vibrational mode. That is, increase in bond strength leads to corresponding frequency increase and vice versa.

Uses and Applications

Infrared spectroscopy is a simple and reliable technique widely used in both organic and inorganic chemistry, in research and industry. It is used in quality control, dynamic measurement, and monitoring applications such as the long-term unattended measurement of CO_2 concentrations in greenhouses and growth chambers by infrared gas analyzers.

It is also used in forensic analysis in both criminal and civil cases, for example in identifying polymer degradation. It can be used in determining the blood alcohol content of a suspected drunk driver.

IR-spectroscopy has been successfully used in analysis and identification of pigments in paintings and other art objects such as illuminated manuscripts.

A useful way of analyzing solid samples without the need for cutting samples uses ATR or attenuated total reflectance spectroscopy. Using this approach, samples are pressed against the face of a single crystal. The infrared radiation passes through the crystal and only interacts with the sample at the interface between the two materials.

With increasing technology in computer filtering and manipulation of the results, samples in solution can now be measured accurately (water produces a broad absorbance across the range of interest, and thus renders the spectra unreadable without this computer treatment).

Some instruments will also automatically tell you what substance is being measured from a store of thousands of reference spectra held in storage.

Infrared spectroscopy is also useful in measuring the degree of polymerization in polymer manufacture. Changes in the character or quantity of a particular bond are assessed by measuring at a specific frequency over time. Modern research instruments can take infrared measurements across the range of interest as frequently as 32 times a second. This can be done whilst simultaneous measurements are made using other techniques. This makes the observations of chemical reactions and processes quicker and more accurate.

Infrared spectroscopy has also been successfully utilized in the field of semiconductor microelectronics: for example, infrared spectroscopy can be applied to semiconductors like silicon, gallium arsenide, gallium nitride, zinc selenide, amorphous silicon, silicon nitride, etc.

Another important application of Infrared Spectroscopy is in the food industry to measure the concentration of various compounds in different food products.

The instruments are now small, and can be transported, even for use in field trials.

Infrared Spectroscopy is also used in gas leak detection devices such as the DP-IR and EyeCGAs. These devices detect hydrocarbon gas leaks in the transportation of natural gas and crude oil.

In February 2014, NASA announced a greatly upgraded database, based on IR spectroscopy, for tracking polycyclic aromatic hydrocarbons (PAHs) in the universe. According to scientists, more than 20% of the carbon in the universe may be associated with PAHs, possible starting materials for the formation of life. PAHs seem to have been formed shortly after the Big Bang, are widespread throughout the universe, and are associated with new stars and exoplanets.

Isotope Effects

The different isotopes in a particular species may exhibit different fine details in infrared spectroscopy. For example, the O–O stretching frequency (in reciprocal centimeters) of oxyhemocyanin is experimentally determined to be 832 and 788 cm^{-1} for $v(^{16}O-^{16}O)$ and $v(^{18}O-^{18}O)$, respectively.

By considering the O–O bond as a spring, the wavenumber of absorbance, v can be calculated:

$$v = \frac{1}{2\pi c}\sqrt{\frac{k}{\mu}}$$

where k is the spring constant for the bond, c is the speed of light, and μ is the reduced mass of the A–B system:

$$\mu = \frac{m_A m_B}{m_A + m_B}$$

(m_i is the mass of atom i).

The reduced masses for $^{16}O-^{16}O$ and $^{18}O-^{18}O$ can be approximated as 8 and 9 respectively. Thus

$$\frac{v(^{16}O)}{v(^{18}O)} = \sqrt{\frac{9}{8}} \approx \frac{832}{788}.$$

Where v is the wavenumber; [wavenumber = frequency/(speed of light)]

The effect of isotopes, both on the vibration and the decay dynamics, has been found to be stronger than previously thought. In some systems, such as silicon and germanium, the decay of the anti-symmetric stretch mode of interstitial oxygen involves the symmetric stretch mode with a strong isotope dependence. For example, it was shown that for a natural silicon sample, the lifetime of the anti-symmetric vibration is 11.4 ps. When the isotope of one of the silicon atoms is increased to ^{29}Si, the lifetime increases to 19 ps. In similar manner, when the silicon atom is changed to ^{30}Si, the lifetime becomes 27 ps.

Two-dimensional IR

Two-dimensional infrared correlation spectroscopy analysis combines multiple samples of infrared spectra to reveal more complex properties. By extending the spectral information of a perturbed sample, spectral analysis is simplified and resolution is enhanced. The 2D synchronous and 2D asynchronous spectra represent a graphical overview of the spectral changes due to a perturbation (such as a changing concentration or changing temperature) as well as the relationship between the spectral changes at two different wavenumbers.

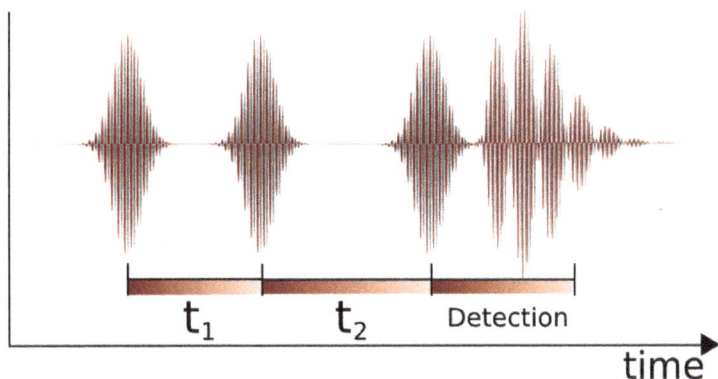

Pulse Sequence used to obtain a two-dimensional Fourier transform infrared spectrum. The time period τ_1 is usually referred to as the coherence time and the second time period is τ_2 known as the waiting time. The excitation frequency is obtained by Fourier transforming along the τ_1 axis.

Nonlinear two-dimensional infrared spectroscopy is the infrared version of correlation spectroscopy. Nonlinear two-dimensional infrared spectroscopy is a technique that has become available with the development of femtosecond infrared laser pulses. In this experiment, first a set of pump pulses is applied to the sample. This is followed by a waiting time during which the system is allowed to relax. The typical waiting time lasts from zero to several picoseconds, and the duration can be controlled with a resolution of tens of femtoseconds. A probe pulse is then applied, resulting in the emission of a signal from the sample. The nonlinear two-dimensional infrared spectrum is a two-dimensional correlation plot of the frequency ω_1 that was excited by the initial pump pulses and the frequency ω_3 excited by the probe pulse after the waiting time. This allows the observation of coupling between different vibrational modes; because of its extremely fine time resolution, it can be used to monitor molecular dynamics on a picosecond timescale. It is still a largely unexplored technique and is becoming increasingly popular for fundamental research.

As with two-dimensional nuclear magnetic resonance (2DNMR) spectroscopy, this technique spreads the spectrum in two dimensions and allows for the observation of cross peaks that contain information on the coupling between different modes. In contrast to 2DNMR, nonlinear two-dimensional infrared spectroscopy also involves the excitation to overtones. These excitations result in excited state absorption peaks located below the diagonal and cross peaks. In 2DNMR, two distinct techniques, COSY and NOESY, are frequently used. The cross peaks in the first are related to the scalar coupling, while in the latter they are related to the spin transfer between different nuclei. In nonlinear two-dimensional infrared spectroscopy, analogs have been drawn to these 2DNMR techniques. Nonlinear two-dimensional infrared spectroscopy with zero waiting time corresponds to COSY, and nonlinear two-dimensional infrared spectroscopy with finite waiting time allowing vibrational population transfer corresponds to NOESY. The COSY variant of nonlinear two-dimensional infrared spectroscopy has been used for determination of the secondary structure content of proteins.

Aim:

To determine the activity of the enzyme alkaline phosphatase.

Introduction:

Enzymes play essential roles by carrying out a plethora of biological reactions. Just because a reaction has very large negative free energy change does not imply that reaction will take place at rapid rate. What it implies is that the $\dfrac{[product]}{[substrate]}$ concentration ratio is smaller than that at equilibrium. Oxidation of glucose into CO_2 and H_2O, for example, is a reaction with $\Delta G'$ of $-686\ kcal/mol$. The glucose, therefore, is thermodynamically unstable. But we know, by experience, that a glucose solution does not break down into CO_2 and H_2O at a measurable rate. We can say that glucose is kinetically stable. The kinetic stability is provided by the large energy barrier between the reactant and the product.

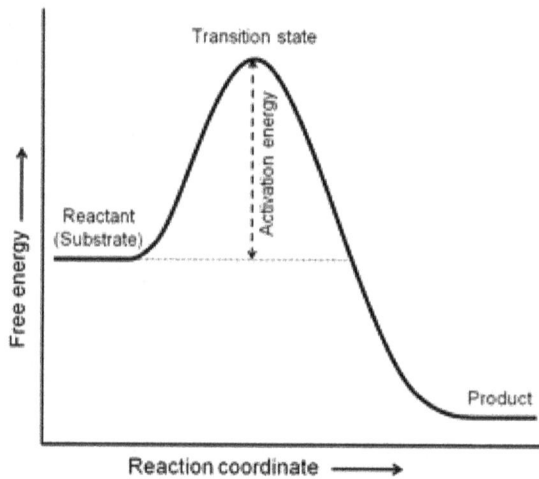

A diagrammatic representation of free energies of reactant,
transition state, and the product

As is clear from the figure, the reactants need excess energy, the activation energy (E_a) to cross the energy barrier between reactants and the products. The rate of the reaction is determined by the number of molecules that enter the transition state per unit time. The number of molecules populating the transition state can be increased either by increasing the temperature or by somehow decreasing the activation energy. As biological organisms survive and function within a narrow temperature window, they can't increase the rate of reaction by increasing the temperature. They manage to carry out a plethora of chemical reactions by means of enzymes that function as biological catalysts by decreasing the activation energy. The enzymes can enhance the reaction rates by up to 15 orders of magnitude. It is important to note that the enzymes do not change the equilibrium constant (K_{eq}) or free energy change (ΔG) of the reaction. Each enzyme present in a cell has its characteristic enzyme parameters. The plot of initial reaction velocity, V_0 against the substrate concentration $[S]$ has same general shape (rectangular hyperbolic shape) which is given by Michaelis-Menten equation:

$$V_0 = \frac{V_{max}[s]}{K_m + [s]}$$

where, V_0 is the initial reaction rate, V_{max} is the maximum rate, $[S]$ is the molar substrate concentration, and K_m is a constant called Michaelis constant.

V_{max} and K_m are the characteristic properties of an enzyme. As is clear from the equation, K_m can be defined as the substrate concentration at which initial reaction rate,

$$V_0 \text{ equals } \frac{V_{max}}{2}.$$

The response of enzymes to the concentrations of substrates and products plays important role in the reaction control. This behavior of enzymes to the substrate/product concentration is studied under enzyme kinetics and is used to determine the important enzyme parameters such as K_m and V_{max}. We have chosen to study the kinetics of the enzyme alkaline phosphatase. The enzyme catalyses the hydrolysis of a phosphoester bond, producing inorganic phosphate (P_i) and an alcohol. We have chosen p-nitrophenylphosphate as the substrate for the hydrolysis reaction. Para-nitrophenylphosphate is a colourless compound; the enzyme, alkaline phosphatase hydrolyses the phosphoester bond to produce the coloured product, p-nitrophenol which can be detected colorimetrically.

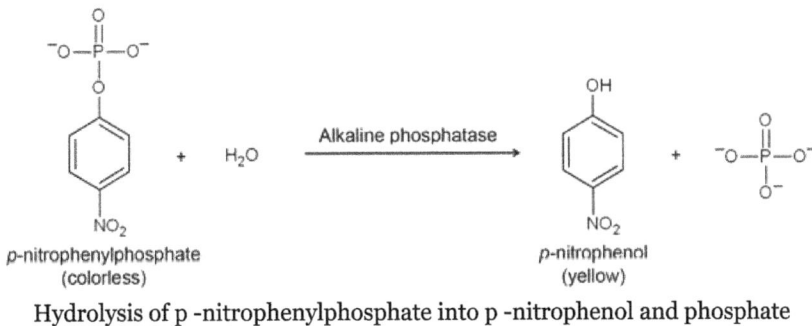

Hydrolysis of p -nitrophenylphosphate into p -nitrophenol and phosphate

Materials:

Equipments:

1. UV/Visible spectrophotometer

2. Weighing balance

Reagents:

1. 100 mM Tris-HCl buffer, pH 8.0

2. Para-nitrophenol (PNP)

3. Para-nitrophenylphosphate (PNPP)

4. Alkaline phosphatase from *E. coli*

Glassware, plasticware, etc.:

1. 1.5 *ml* microfuge tubes

2. Pipettes

3. Pipette tips

4. A pair of matched glass or quartz cuvettes (volume: 3 *ml*)

Procedure:

Standard curve of PNP

Table: Observation table for the enzyme assay

Tube No.	Tris-HCl buffer (*ml*)	PNP solution (*ml*)	PNP concentration (*μM*)	A_{410}
1	2.7	0.3	10	
2	2.4	0.6	20	
3	2.1	0.9	30	
4	1.8	1.2	50	
5	1.5	1.5	50	
6	1.2	1.8	60	
7	0.9	2.1	70	
8	0.6	2.4	80	
9	0.3	2.7	90	
10	0	3.0	100	
11	6.0	Blank (0)	Zero	

1. Switch ON the spectrophotometer and allow it 30 *min* warm up.

2. Meanwhile, prepare 0.1 *mM* PNP solution in 100 *mM* Tris-HCl buffer, pH 8.0

3. Take 11 microfuge tubes and label them from 1 – 11.

4. Prepare the PNP dilutions as shown in the table.

5. Set the spectrophotometer to 410 nm and select the "Absorbance mode".

6. Use the blank (Tube No. 11) in both the cuvettes to set the spectrophotometer readings to ZERO.

7. Measure the absorbance of tubes 1 – 10 against the blank and record the readings in the table.

8. Plot the absorbance values against the PNP concentration.

9. Fit the data points using linear regression to obtain the standard curve.

Enzyme kinetics and determination of K_m and V_{max}

1. Prepare 100 μM solution of the enzyme (200 μl) in the Tris-HCl buffer, pH 8.0.

2. Prepare 5, 10, 15, 20, 25, 50, 75, and 100 mM PNPP solutions in Tris-HCl buffer, pH 8.0.

3. Record the absorbance at 410 nm using each of the PNPP solutions as follows:

    a. Add 2.97 ml Tris-HCl buffer in each of the 3 ml cuvettes.

    b. Add 30 μl of PNPP solution in both the cuvettes and mix well.

    c. ZERO the reading at 410 nm.

    d. Add 20 μl of the enzyme solution to the cuvette kept in sample cell, start the stop watch, cover the cuvette with a piece of parafilm, and quickly mix the contents by 4-5 inversions.

    e. Immediately record the absorbance and then after 10 seconds interval for 2 min.

4. Repeat the assay for each of the samples at least once and take the average readings for analysis.

5. Plot the average absorbance value against time for each of the samples. This gives the time course of the enzymatic reaction.

6. Calculate the initial velocity, V_o for each of the substrate (PNPP) concentration.

    a. The plot between absorbance against time is linear for the initial part of the plot and V_o is simply the slope of this line

    b. Fit the initial region of the curve (first 3 or 4 points) linearly and determine the slope of the line.

7. Plot V_o against substrate concentration to obtain the Michaelis-Menten curve.

8. The Michaelis-Meneten equation shown in the equation can be rewritten as:

$$\frac{1}{V} = \frac{K_m}{V_{max}} \frac{1}{[s]} + \frac{1}{V_{max}}$$

A plot between $\frac{1}{v_0}$ and $\frac{1}{[s]}$ gives a straight line with a slope $\frac{K_m}{V_{max}}$ and an intercept of $\frac{1}{V_{max}}$ on $\frac{1}{V_0}$ axis. This plot is known as Lineweaver-Burk plot or double-reciprocal plot and allows easy determination of the K_m and V_{max} of the enzyme.

9. Calculate the $\frac{1}{V_0}$ and $\frac{1}{[s]}$ for each of the substrate concentration, obtain the Lineweaver-Burk plot and calculate the K_m and V_{max} from the plot.

Electrophoresis

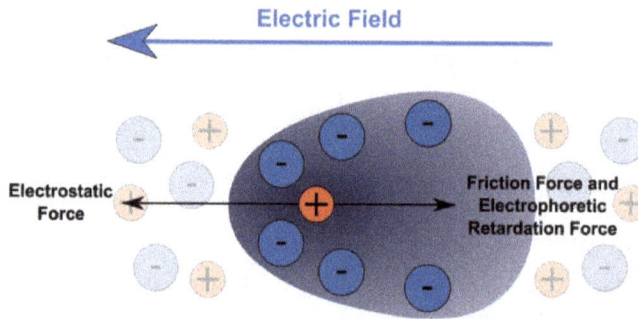

Illustration of electrophoresis

Electrophoresis is the motion of dispersed particles relative to a fluid under the influence of a spatially uniform electric field. This electrokinetic phenomenon was observed for the first time in 1807 by Ferdinand Frederic Reuss (Moscow State University), who noticed that the application of a constant electric field caused clay particles dispersed in water to migrate. It is ultimately caused by the presence of a charged interface between the particle surface and the surrounding fluid. It is the basis for a number of analytical techniques used in chemistry for separating molecules by size, charge, or binding affinity.

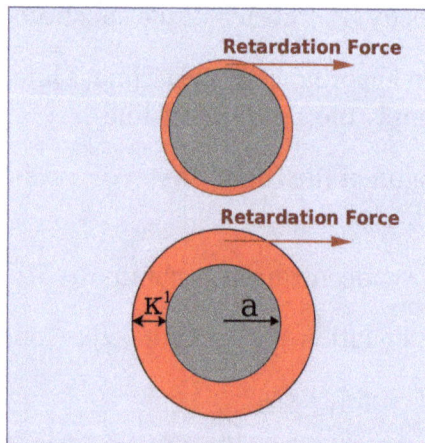

Illustration of electrophoresis retardation

Electrophoresis of positively charged particles (cations) is called cataphoresis, while electrophoresis of negatively charged particles (anions) is called anaphoresis. Electrophoresis is a technique used in laboratories in order to separate macromolecules based on size. The technique applies a negative charge so proteins move towards a positive charge. This is used for both DNA and RNA analysis. Polyacrylamide gel electrophoresis (PAGE) has a clearer resolution than agarose and is more suitable for quantitative analysis. In this technique DNA foot-printing can identify how proteins bind to DNA. It can be used to separate proteins by size, density and purity. It can also be used for plasmid analysis, which develops our understanding of bacteria becoming resistant to antibiotics.

Theory

Suspended particles have an electric surface charge, strongly affected by surface adsorbed species, on which an external electric field exerts an electrostatic Coulomb force. According to the double layer theory, all surface charges in fluids are screened by a diffuse layer of ions, which has the same absolute charge but opposite sign with respect to that of the surface charge. The electric field also exerts a force on the ions in the diffuse layer which has direction opposite to that acting on the surface charge. This latter force is not actually applied to the particle, but to the ions in the diffuse layer located at some distance from the particle surface, and part of it is transferred all the way to the particle surface through viscous stress. This part of the force is also called electrophoretic retardation force. When the electric field is applied and the charged particle to be analyzed is at steady movement through the diffuse layer, the total resulting force is zero :

$$F_{tot} = 0 = F_{el} + F_f + F_{ret}$$

Considering the drag on the moving particles due to the viscosity of the dispersant, in the case of low Reynolds number and moderate electric field strength E, the drift velocity of a dispersed particle v is simply proportional to the applied field, which leaves the electrophoretic mobility μ_e defined as:

$$\mu_e = \frac{v}{E}.$$

The most well known and widely used theory of electrophoresis was developed in 1903 by Smoluchowski:

$$\mu_e = \frac{\varepsilon_r \varepsilon_0 \zeta}{\eta},$$

where ε_r is the dielectric constant of the dispersion medium, ε_0 is the permittivity of free space ($C^2\,N^{-1}\,m^{-2}$), η is dynamic viscosity of the dispersion medium (Pa s), and ζ

is zeta potential (i.e., the electrokinetic potential of the slipping plane in the double layer).

The Smoluchowski theory is very powerful because it works for dispersed particles of any shape at any concentration. It has limitations on its validity. It follows, for instance, because it does not include Debye length κ^{-1}. However, Debye length must be important for electrophoresis, as follows immediately from the Figure on the right. Increasing thickness of the double layer (DL) leads to removing the point of retardation force further from the particle surface. The thicker the DL, the smaller the retardation force must be.

Detailed theoretical analysis proved that the Smoluchowski theory is valid only for sufficiently thin DL, when particle radius a is much greater than the Debye length:

$$a\kappa \gg 1.$$

This model of "thin double layer" offers tremendous simplifications not only for electrophoresis theory but for many other electrokinetic theories. This model is valid for most aqueous systems, where the Debye length is usually only a few nanometers. It only breaks for nano-colloids in solution with ionic strength close to water.

The Smoluchowski theory also neglects the contributions from surface conductivity. This is expressed in modern theory as condition of small Dukhin number:

$$Du \ll 1$$

In the effort of expanding the range of validity of electrophoretic theories, the opposite asymptotic case was considered, when Debye length is larger than particle radius:

$$a\kappa < 1.$$

Under this condition of a "thick double layer", Hückel predicted the following relation for electrophoretic mobility:

$$\mu_e = \frac{2\varepsilon_r\varepsilon_0\zeta}{3\eta}.$$

This model can be useful for some nanoparticles and non-polar fluids, where Debye length is much larger than in the usual cases.

There are several analytical theories that incorporate surface conductivity and eliminate the restriction of a small Dukhin number, pioneered by Overbeek. and Booth. Modern, rigorous theories valid for any Zeta potential and often any $a\kappa$ stem mostly from Dukhin–Semenikhin theory. In the thin double layer limit, these theories confirm the numerical solution to the problem provided by O'Brien and White.

Polyacrylamide Gel Electrophoresis

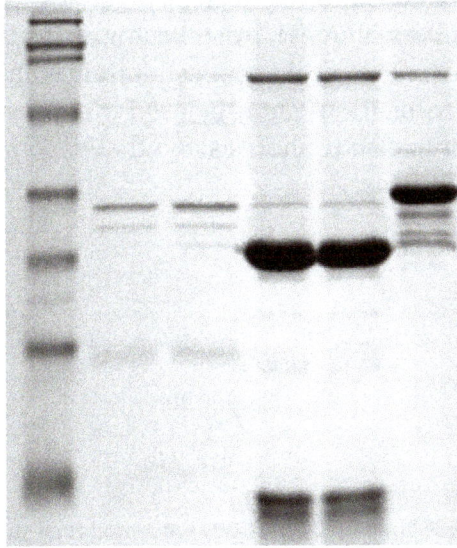

Picture of an SDS-PAGE. The molecular markers (ladder) are in the left lane

Polyacrylamide gel electrophoresis (PAGE), describes a technique widely used in biochemistry, forensics, genetics, molecular biology and biotechnology to separate biological macromolecules, usually proteins or nucleic acids, according to their electrophoretic mobility. Mobility is a function of the length, conformation and charge of the molecule.

As with all forms of gel electrophoresis, molecules may be run in their native state, preserving the molecules' higher-order structure, or a chemical denaturant may be added to remove this structure and turn the molecule into an unstructured linear chain whose mobility depends only on its length and mass-to-charge ratio. For nucleic acids, urea is the most commonly used denaturant. For proteins, sodium dodecyl sulfate (SDS) is an anionic detergent applied to protein samples to linearize proteins and to impart a negative charge to linearized proteins. This procedure is called SDS-PAGE. In most proteins, the binding of SDS to the polypeptide chain imparts an even distribution of charge per unit mass, thereby resulting in a fractionation by approximate size during electrophoresis. Proteins that have a greater hydrophobic content, for instance many membrane proteins, and those that interact with surfactants in their native environment, are intrinsically harder to treat accurately using this method, due to the greater variability in the ratio of bound SDS.

Procedure

Sample Preparation

Samples may be any material containing proteins or nucleic acids. These may be biologically derived, for example from prokaryotic or eukaryotic cells, tissues, viruses, environmental samples, or purified proteins. In the case of solid tissues or cells, these are often first broken down mechanically using a blender (for larger sample volumes), using a homogenizer (smaller volumes), by sonicator or by using cycling of high pressure, and a combination of biochemical and mechanical techniques – including various types of filtration and centrifugation – may be used to separate different cell compartments and organelles prior to electrophoresis. Synthetic biomolecules such as oligonucleotides may also be used as analytes.

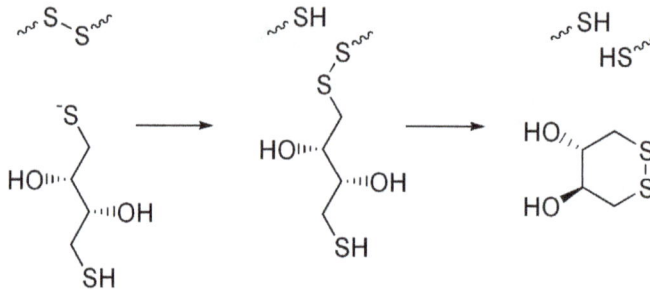

Reduction of a typical disulfide bond by DTT via two sequential thiol-disulfide exchange reactions.

The sample to analyze is optionally mixed with a chemical denaturant if so desired, usually SDS for proteins or urea for nucleic acids. SDS is an anionic detergent that denatures secondary and non–disulfide–linked tertiary structures, and additionally applies a negative charge to each protein in proportion to its mass. Urea breaks the hydrogen bonds between the base pairs of the nucleic acid, causing the constituent strands to separate. Heating the samples to at least 60°C further promotes denaturation.

In addition to SDS, proteins may optionally be briefly heated to near boiling in the presence of a reducing agent, such as dithiothreitol (DTT) or 2-mercaptoethanol (beta-mercaptoethanol/BME), which further denatures the proteins by reducing disulfide linkages, thus overcoming some forms of tertiary protein folding, and breaking up quaternary protein structure (oligomeric subunits). This is known as reducing SDS-PAGE.

A tracking dye may be added to the solution. This typically has a higher electrophoretic mobility than the analytes to allow the experimenter to track the progress of the solution through the gel during the electrophoretic run.

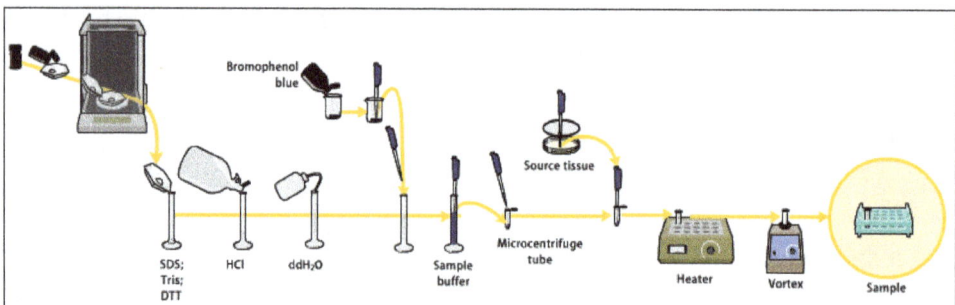

Preparing Acrylamide Gels

The gels typically consist of acrylamide, bisacrylamide, the optional denaturant (SDS or urea), and a buffer with an adjusted pH. The solution may be degassed under a vacuum to prevent the formation of air bubbles during polymerization. Alternatively, butanol may be added to the resolving gel (for proteins) after it is poured, as butanol removes bubbles and makes the surface smooth. A source of free radicals and a stabilizer, such as ammonium persulfate and TEMED are added to initiate polymerization. The polymerization reaction creates a gel because of the added bisacrylamide, which can form cross-links between two acrylamide molecules. The ratio of bisacrylamide to acrylamide can be varied for special purposes, but is generally about 1 part in 35. The acrylamide concentration of the gel can also be varied, generally in the range from 5% to 25%. Lower percentage gels are better for resolving very high molecular weight molecules, while much higher percentages are needed to resolve smaller proteins.

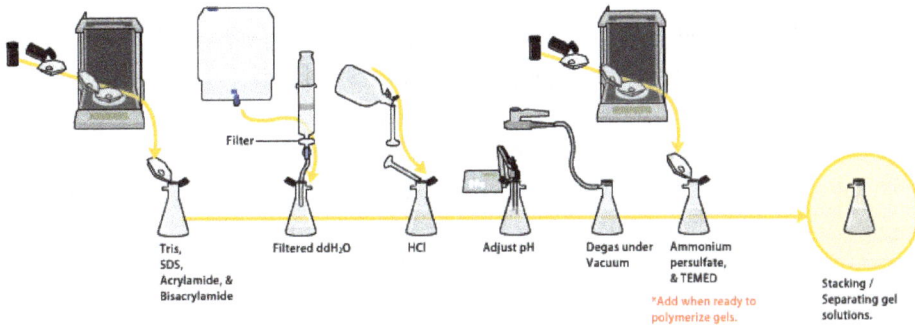

Tris, SDS, Acrylamide, & Bisacrylamide   Filtered ddH₂O   HCl   Adjust pH   Degas under Vacuum   Ammonium persulfate, & TEMED   *Add when ready to polymerize gels.   Stacking / Separating gel solutions.

Gels are usually polymerized between two glass plates in a gel caster, with a comb inserted at the top to create the sample wells. After the gel is polymerized the comb can be removed and the gel is ready for electrophoresis.

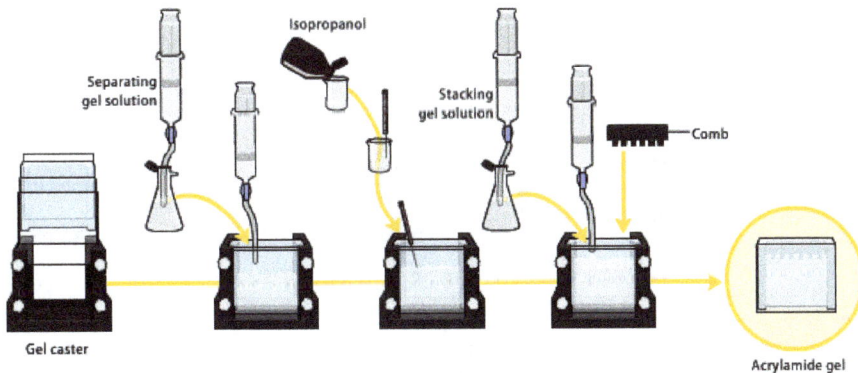

Gel caster   Separating gel solution   Isopropanol   Stacking gel solution   Comb   Acrylamide gel

Various buffer systems are used in PAGE depending on the nature of the sample and the experimental objective. The buffers used at the anode and cathode may be the same or different.

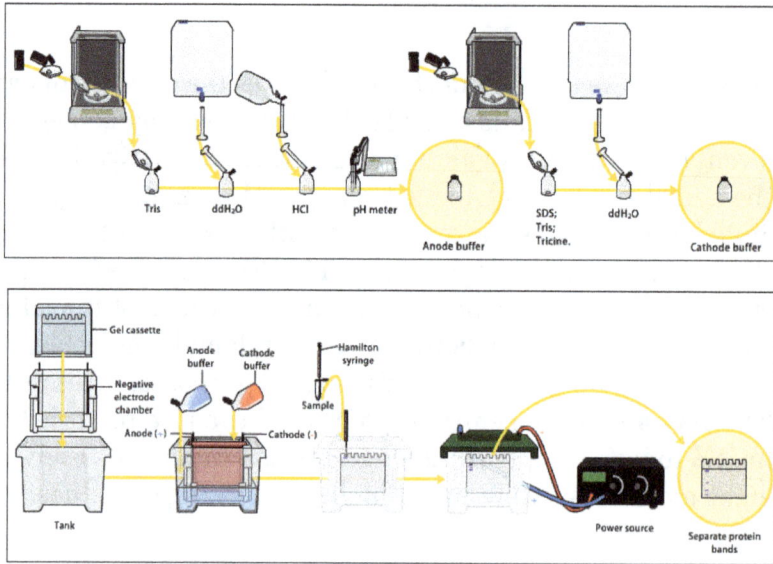

An electric field is applied across the gel, causing the negatively charged proteins or nucleic acids to migrate across the gel away from the negative electrode (which is the cathode being that this is an electrolytic rather than galvanic cell) and towards the positive electrode (the anode). Depending on their size, each biomolecule moves differently through the gel matrix: small molecules more easily fit through the pores in the gel, while larger ones have more difficulty. The gel is run usually for a few hours, though this depends on the voltage applied across the gel; migration occurs more quickly at higher voltages, but these results are typically less accurate than at those at lower voltages. After the set amount of time, the biomolecules have migrated different distances based on their size. Smaller biomolecules travel farther down the gel, while larger ones remain closer to the point of origin. Biomolecules may therefore be separated roughly according to size, which depends mainly on molecular weight under denaturing conditions, but also depends on higher-order conformation under native conditions. However, certain glycoproteins behave anomalously on SDS gels.

Further Processing

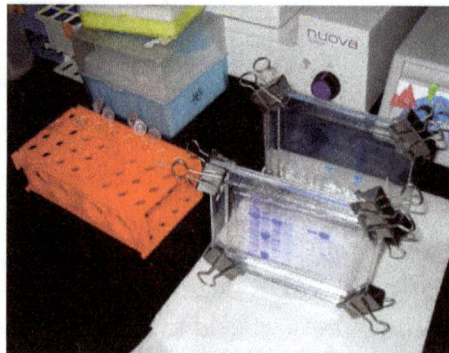

Two SDS-PAGE-gels after a completed run

Following electrophoresis, the gel may be stained (for proteins, most commonly with Coomassie Brilliant Blue R-250; for nucleic acids, ethidium bromide; or for either, silver stain), allowing visualization of the separated proteins, or processed further (e.g. Western blot). After staining, different species biomolecules appear as distinct bands within the gel. It is common to run molecular weight size markers of known molecular weight in a separate lane in the gel to calibrate the gel and determine the approximate molecular mass of unknown biomolecules by comparing the distance traveled relative to the marker.

For proteins, SDS-PAGE is usually the first choice as an assay of purity due to its reliability and ease. The presence of SDS and the denaturing step make proteins separate, approximately based on size, but aberrant migration of some proteins may occur. Different proteins may also stain differently, which interferes with quantification by staining. PAGE may also be used as a preparative technique for the purification of proteins. For example, quantitative preparative native continuous polyacrylamide gel electrophoresis (QPNC-PAGE) is a method for separating native metalloproteins in complex biological matrices.

Chemical Ingredients and their Roles

Polyacrylamide gel (PAG) had been known as a potential embedding medium for sectioning tissues as early as 1964, and two independent groups employed PAG in electrophoresis in 1959. It possesses several electrophoretically desirable features that make it a versatile medium. It is a synthetic, thermo-stable, transparent, strong, chemically relatively inert gel, and can be prepared with a wide range of average pore sizes. The pore size of a gel is determined by two factors, the total amount of acrylamide present (%T) (T = Total concentration of acrylamide and bisacrylamide monomer) and the amount of cross-linker (%C) (C = bisacrylamide concentration). Pore size decreases with increasing %T; with cross-linking, 5%C gives the smallest pore size. Any increase or decrease in %C from 5% increases the pore size, as pore size with respect to %C is a parabolic function with vertex as 5%C. This appears to be because of non-homogeneous bundling of polymer strands within the gel. This gel material can also withstand high voltage gradients, is amenable to various staining and destaining procedures, and can be digested to extract separated fractions or dried for autoradiography and permanent recording.

Components

- Chemical buffer Stabilizes the pH value to the desired value within the gel itself and in the electrophoresis buffer. The choice of buffer also affects the electrophoretic mobility of the buffer counterions and thereby the resolution of the gel. The buffer should also be unreactive and not modify or react with most proteins. Different buffers may be used as cathode and anode buffers, respectively, depending on the application. Multiple pH values may be used within a single gel, for example in DISC electrophoresis. Common buffers in PAGE include Tris, Bis-Tris, or imidazole.

- Counterion balance the intrinsic charge of the buffer ion and also affect the electric field strength during electrophoresis. Highly charged and mobile ions are often avoided in SDS-PAGE cathode buffers, but may be included in the gel itself, where it migrates ahead of the protein. In applications such as DISC SDS-PAGE the pH values within the gel may vary to change the average charge of the counterions during the run to improve resolution. Popular counterions are glycine and tricine. Glycine has been used as the source of trailing ion or slow ion because its pKa is 9.69 and mobility of glycinate are such that the effective mobility can be set at a value below that of the slowest known proteins of net negative charge in the pH range. The minimum pH of this range is approximately 8.0.

- Acrylamide (C_3H_5NO; mW: 71.08). When dissolved in water, slow, spontaneous autopolymerization of acrylamide takes place, joining molecules together by head on tail fashion to form long single-chain polymers. The presence of a free radical-generating system greatly accelerates polymerization. This kind of reaction is known as Vinyl addition polymerisation. A solution of these polymer chains becomes viscous but does not form a gel, because the chains simply slide over one another. Gel formation requires linking various chains together. Acrylamide is carcinogenic, a neurotoxin, and a reproductive toxin. It is also essential to store acrylamide in a cool dark and dry place to reduce autopolymerisation and hydrolysis.

- Bisacrylamide (*N,N'*-Methylenebisacrylamide) ($C_7H_{10}N_2O_2$; mW: 154.17). Bisacrylamide is the most frequently used cross linking agent for polyacrylamide gels. Chemically it can be thought of as two acrylamide molecules coupled head to head at their non-reactive ends. Bisacrylamide can crosslink two polyacrylamide chains to one another, thereby resulting in a gel.

- Sodium Dodecyl Sulfate (SDS) ($C_{12}H_{25}NaO_4S$; mW: 288.38). (only used in denaturing protein gels) SDS is a strong detergent agent used to denature native proteins to unfolded, individual polypeptides. When a protein mixture is heated to 100°C in presence of SDS, the detergent wraps around the polypeptide backbone. It binds to polypeptides in a constant weight ratio of 1.4 g SDS/g of polypeptide. In this process, the intrinsic charges of polypeptides become negligible when compared to the negative charges contributed by SDS. Thus polypeptides after treatment become rod-like structures possessing a uniform charge density, that is same net negative charge per unit weight. The electrophoretic mobilities of these proteins is a linear function of the logarithms of their molecular weights.

Without SDS, different proteins with similar molecular weights would migrate differently due to differences in mass-charge ratio, as each protein has an isoelectric point and molecular weight particular to its primary structure. This is known as Native PAGE. Adding SDS solves this problem, as it binds to and unfolds the protein, giving a near uniform negative charge along the length of the polypeptide.

- Urea ($CO(NH_2)_2$; mW: 60.06). Urea is a chaotropic agent that increases the entropy of the system by interfering with intramolecular interactions mediated by non-covalent forces such as hydrogen bonds and van der Waals forces. Macromolecular structure is dependent on the net effect of these forces, therefore it follows that an increase in chaotropic solutes denatures macromolecules.

- Ammonium persulfate (APS) ($N_2H_8S_2O_8$; mW: 228.2). APS is a source of free radicals and is often used as an initiator for gel formation. An alternative source of free radicals is riboflavin, which generated free radicals in a photochemical reaction.

- TEMED (*N, N, N', N'*-tetramethylethylenediamine) ($C_6H_{16}N_2$; mW: 116.21). TEMED stabilizes free radicals and improves polymerization. The rate of polymerisation and the properties of the resulting gel depend on the concentrations of free radicals. Increasing the amount of free radicals results in a decrease in the average polymer chain length, an increase in gel turbidity and a decrease in gel elasticity. Decreasing the amount shows the reverse effect. The lowest catalytic concentrations that allow polymerisation in a reasonable period of time should be used. APS and TEMED are typically used at approximately equimolar concentrations in the range of 1 to 10 mM.

Chemicals for Processing and Visualization

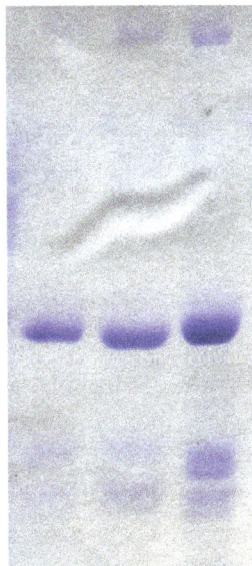

PAGE of rotavirus proteins stained with Coomassie blue

The following chemicals and procedures are used for processing of the gel and the protein samples visualized in it:

- Tracking dye. As proteins and nucleic acids are mostly colorless, their progress through the gel during electrophoresis cannot be easily followed. Anionic dyes of a known electrophoretic mobility are therefore usually included in the PAGE sample buffer. A very common tracking dye is Bromophenol blue (BPB, 3',3",5',5" tetrabromophenolsulfonphthalein). This dye is coloured at alkali and neutral pH and is a small negatively charged molecule that moves towards the anode. Being a highly mobile molecule it moves ahead of most proteins. As it reaches the anodic end of the electrophoresis medium electrophoresis is stopped. It can weakly bind to some proteins and impart a blue colour. Other common tracking dyes are xylene cyanol, which has lower mobility, and Orange G, which has a higher mobility.

- Loading aids. Most PAGE systems are loaded from the top into wells within the gel. To ensure that the sample sinks to the bottom of the gel, sample buffer is supplemented with additives that increase the density of the sample. These additives should be non-ionic and non-reactive towards proteins to avoid interfering with electrophoresis. Common additives are glycerol and sucrose.

- Coomassie Brilliant Blue R-250 (CBB)(C$_{45}$H$_{44}$N$_3$NaO$_7$S$_2$; mW: 825.97). CBB is the most popular protein stain. It is an anionic dye, which non-specifically binds to proteins. The structure of CBB is predominantly non-polar, and it is usually used in methanolic solution acidified with acetic acid. Proteins in the gel are fixed by acetic acid and simultaneously stained. The excess dye incorporated into the gel can be removed by destaining with the same solution without the dye. The proteins are detected as blue bands on a clear background. As SDS is also anionic, it may interfere with staining process. Therefore, large volume of staining solution is recommended, at least ten times the volume of the gel.

- Ethidium bromide (EtBr) is the traditionally most popular nucleic acid stain.

- Silver staining. Silver staining is used when more sensitive method for detection is needed, as classical Coomassie Brilliant Blue staining can usually detect a 50 ng protein band, Silver staining increases the sensitivity typically 50 times. The exact chemical mechanism by which this happens is still largely unknown. Silver staining was introduced by Kerenyi and Gallyas as a sensitive procedure to detect trace amounts of proteins in gels. The technique has been extended to the study of other biological macromolecules that have been separated in a variety of supports. Many variables can influence the colour intensity and every protein has its own staining characteristics; clean glassware, pure reagents and water of highest purity are the key points to successful staining. Silver staining was developed in the 14th century for colouring the surface of glass. It has been used extensively for this purpose since the 16th century. The colour produced by the early silver stains ranged between light yellow and an orange-red. Camillo Golgi perfected the silver staining for the study of the nervous system. Golgi's method stains a limited number of cells at random in their entirety.

- Western blotting is a process by which proteins separated in the acrylamide gel are electrophoretically transferred to a stable, manipulable membrane such as a nitrocellulose, nylon, or PVDF membrane. It is then possible to apply immunochemical techniques to visualise the transferred proteins, as well as accurately identify relative increases or decreases of the protein of interest.

Chromatography

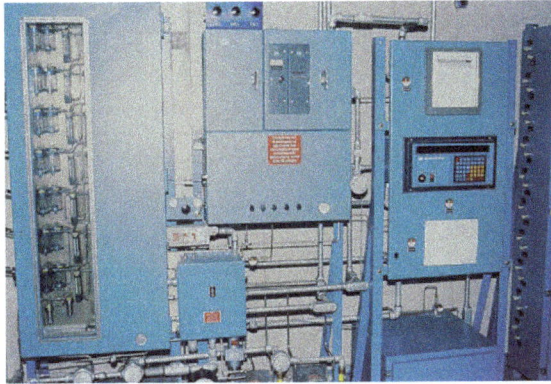

Pictured is a sophisticated gas chromatography system. This instrument records concentrations of acrylonitrile in the air at various points throughout the chemical laboratory.

Automated fraction collector and sampler for chromatographic techniques

Chromatography is a laboratory technique for the separation of a mixture. The mixture is dissolved in a fluid called the *mobile phase,* which carries it through a structure holding another material called the *stationary phase.* The various constituents of the mixture travel at different speeds, causing them to separate. The separation is based on differential partitioning between the mobile and stationary phases. Subtle differences in a compound's partition coefficient result in differential retention on the stationary phase and thus changing the separation.

Chromatography may be preparative or analytical. The purpose of preparative chromatography is to separate the components of a mixture for more advanced use (and is thus a form of purification). Analytical chromatography is done normally with smaller amounts of material and is for measuring the relative proportions (or establishing the presence) of analytes in a mixture. The two are not mutually exclusive.

History

Thin layer chromatography is used to separate components of a plant extract, illustrating the experiment with plant pigments that gave chromatography its name

Chromatography was first employed in Russia by the Italian-born scientist Mikhail Tsvet in 1900. He continued to work with chromatography in the first decade of the 20th century, primarily for the separation of plant pigments such as chlorophyll, carotenes, and xanthophylls. Since these components have different colors (green, orange, and yellow, respectively) they gave the technique its name. New types of chromatography developed during the 1930s and 1940s made the technique useful for many separation processes.

Chromatography technique developed substantially as a result of the work of Archer John Porter Martin and Richard Laurence Millington Synge during the 1940s and 1950s, for which they won a Nobel prize. They established the principles and basic techniques of partition chromatography, and their work encouraged the rapid development of several chromatographic methods: paper chromatography, gas chromatography, and what would become known as high performance liquid chromatography. Since then, the technology has advanced rapidly. Researchers found that the main principles of Tsvet's chromatography could be applied in many different ways, resulting in the different varieties of chromatography described below. Advances are continually improving the technical performance of chromatography, allowing the separation of increasingly similar molecules.

Chromatography Terms

- The analyte is the substance to be separated during chromatography. It is also normally what is needed from the mixture.

- Analytical chromatography is used to determine the existence and possibly also the concentration of analyte(s) in a sample.

- A bonded phase is a stationary phase that is covalently bonded to the support particles or to the inside wall of the column tubing.

- A chromatogram is the visual output of the chromatograph. In the case of an optimal separation, different peaks or patterns on the chromatogram correspond to different components of the separated mixture.

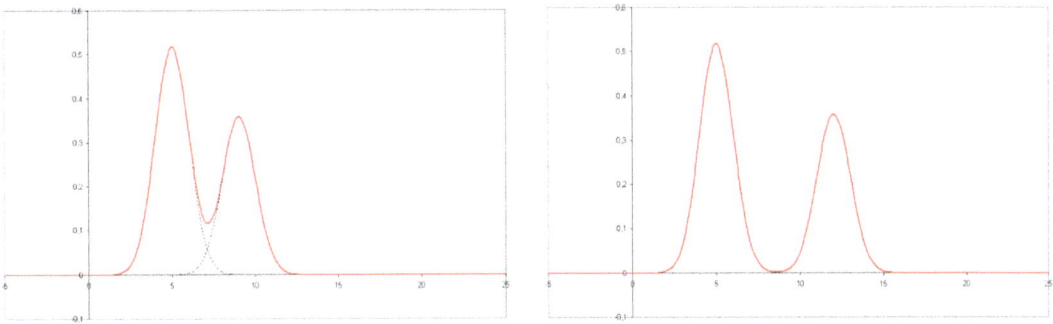

Plotted on the x-axis is the retention time and plotted on the y-axis a signal (for example obtained by a spectrophotometer, mass spectrometer or a variety of other detectors) corresponding to the response created by the analytes exiting the system. In the case of an optimal system the signal is proportional to the concentration of the specific analyte separated.

- A chromatograph is equipment that enables a sophisticated separation, e.g. gas chromatographic or liquid chromatographic separation.

- Chromatography is a physical method of separation that distributes components to separate between two phases, one stationary (stationary phase), the other (the mobile phase) moving in a definite direction.

- The eluate is the mobile phase leaving the column.

- The eluent is the solvent that carries the analyte.

- An eluotropic series is a list of solvents ranked according to their eluting power.

- An immobilized phase is a stationary phase that is immobilized on the support particles, or on the inner wall of the column tubing.

- The mobile phase is the phase that moves in a definite direction. It may be a liquid (LC and Capillary Electrochromatography (CEC)), a gas (GC), or a supercritical fluid (supercritical-fluid chromatography, SFC). The mobile phase consists of the sample being separated/analyzed and the solvent that moves the sample through the column. In the case of HPLC the mobile phase consists of a non-polar solvent(s) such as hexane in normal phase or a polar solvent such as methanol in reverse phase chromatography and the sample being separated. The mobile phase moves through the chromatography column (the stationary phase) where the sample interacts with the stationary phase and is separated.

- Preparative chromatography is used to purify sufficient quantities of a substance for further use, rather than analysis.

- The retention time is the characteristic time it takes for a particular analyte to pass through the system (from the column inlet to the detector) under set conditions.

- The sample is the matter analyzed in chromatography. It may consist of a single component or it may be a mixture of components. When the sample is treated in the course of an analysis, the phase or the phases containing the analytes of interest is/are referred to as the sample whereas everything out of interest separated from the sample before or in the course of the analysis is referred to as waste.

- The solute refers to the sample components in partition chromatography.

- The solvent refers to any substance capable of solubilizing another substance, and especially the liquid mobile phase in liquid chromatography.

- The stationary phase is the substance fixed in place for the chromatography procedure. Examples include the silica layer in thin layer chromatography.

- The detector refers to the instrument used for qualitative and quantitative detection of analytes after separation.

Chromatography is based on the concept of partition coefficient. Any solute partitions between two immiscible solvents. When we make one solvent immobile (by adsorption on a solid support matrix) and another mobile it results in most common applications of chromatography. If the matrix support, or stationary phase, is polar (e.g. paper, silica etc.) it is forward phase chromatography, and if it is non-polar (C-18) it is reverse phase.

Techniques by Chromatographic Bed Shape

Column Chromatography

Column chromatography is a separation technique in which the stationary bed is within a tube. The particles of the solid stationary phase or the support coated with a liquid stationary phase may fill the whole inside volume of the tube (packed column) or be concentrated on or along the inside tube wall leaving an open, unrestricted path for the mobile phase in the middle part of the tube (open tubular column). Differences in rates of movement through the medium are calculated to different retention times of the sample.

In 1978, W. Clark Still introduced a modified version of column chromatography called flash column chromatography (flash). The technique is very similar to the traditional column chromatography, except for that the solvent is driven through the column by applying positive pressure. This allowed most separations to be performed in less than 20 minutes, with improved separations compared to the old method. Modern flash chromatography systems are sold as pre-packed plastic cartridges, and the solvent is pumped through the cartridge. Systems may also be linked with detectors and fraction collectors providing automation. The introduction of gradient pumps resulted in quicker separations and less solvent usage.

In expanded bed adsorption, a fluidized bed is used, rather than a solid phase made by a packed bed. This allows omission of initial clearing steps such as centrifugation and filtration, for culture broths or slurries of broken cells.

Phosphocellulose chromatography utilizes the binding affinity of many DNA-binding proteins for phosphocellulose. The stronger a protein's interaction with DNA, the higher the salt concentration needed to elute that protein.

Planar Chromatography

Planar chromatography is a separation technique in which the stationary phase is present as or on a plane. The plane can be a paper, serving as such or impregnated by a substance as the stationary bed (paper chromatography) or a layer of solid particles spread on a support such as a glass plate (thin layer chromatography). Different compounds in the sample mixture travel different distances according to how strongly they interact with the stationary phase as compared to the mobile phase. The specific Retention factor (R_f) of each chemical can be used to aid in the identification of an unknown substance.

Paper Chromatography

Paper chromatography is a technique that involves placing a small dot or line of sample solution onto a strip of *chromatography paper*. The paper is placed in a container with a shallow layer of solvent and sealed. As the solvent rises through the paper, it meets the sample mixture, which starts to travel up the paper with the solvent. This paper is made of cellulose, a polar substance, and the compounds within the mixture travel farther if they are non-polar. More polar substances bond with the cellulose paper more quickly, and therefore do not travel as far.

Thin Layer Chromatography (TLC)

Thin layer chromatography (TLC) is a widely employed laboratory technique and is similar to paper chromatography. However, instead of using a stationary phase of paper, it involves a stationary phase of a thin layer of adsorbent like silica gel, alumina, or cellulose on a flat, inert substrate. Compared to paper, it has the advantage of faster runs, better separations, and the choice between different adsorbents. For even better resolution and to allow for quantification, high-performance TLC can be used. An older popular use had been to differentiate chromosomes by observing distance in gel (separation of was a separate step).

Displacement Chromatography

The basic principle of displacement chromatography is: A molecule with a high affinity for the chromatography matrix (the displacer) competes effectively for binding sites, and thus displace all molecules with lesser affinities. There are distinct differences between displacement and elution chromatography. In elution mode, substances typically emerge from a column in narrow, Gaussian peaks. Wide separation of peaks, preferably to baseline, is desired for maximum purification. The speed at which any component of a mixture travels down the column in elution mode depends on many factors. But for two substances to travel at different speeds, and thereby be resolved, there must be substantial differences in some interaction between the biomolecules and the chromatography matrix. Operating parameters are adjusted to maximize the effect of this difference. In many cases, baseline separation of the peaks can be achieved only with gradient elution and low column loadings. Thus, two drawbacks to elution mode chromatography, especially at the preparative scale, are operational complexity, due to gradient solvent pumping, and low throughput, due to low column loadings. Displacement chromatography has advantages over elution chromatography in that components are resolved into consecutive zones of pure substances rather than "peaks". Because the process takes advantage of the nonlinearity of the isotherms, a larger column feed can be separated on a given column with the purified components recovered at significantly higher concentrations.

Techniques by Physical State of Mobile Phase

Gas Chromatography

Gas chromatography (GC), also sometimes known as gas-liquid chromatography, (GLC), is a separation technique in which the mobile phase is a gas. Gas chromatographic separation is always carried out in a column, which is typically "packed" or "capillary". Packed columns are the routine work horses of gas chromatography, being cheaper and easier to use and often giving adequate performance. Capillary columns generally give far superior resolution and although more expensive are becoming widely used, especially for complex mixtures. Both types of column are made from non-adsorbent and chemically inert materials. Stainless steel and glass are the usual materials for packed columns and quartz or fused silica for capillary columns.

Gas chromatography is based on a partition equilibrium of analyte between a solid or viscous liquid stationary phase (often a liquid silicone-based material) and a mobile gas (most often helium). The stationary phase is adhered to the inside of a small-diameter (commonly 0.53 – 0.18mm inside diameter) glass or fused-silica tube (a capillary column) or a solid matrix inside a larger metal tube (a packed column). It is widely used in analytical chemistry; though the high temperatures used in GC make it unsuitable for high molecular weight biopolymers or proteins (heat denatures them), frequently encountered in biochemistry, it is well suited for use in the petrochemical, environmental monitoring and remediation, and industrial chemical fields. It is also used extensively in chemistry research.

Liquid Chromatography

Liquid chromatography (LC) is a separation technique in which the mobile phase is a liquid. It can be carried out either in a column or a plane. Present day liquid chromatography that generally utilizes very small packing particles and a relatively high pressure is referred to as high performance liquid chromatography (HPLC).

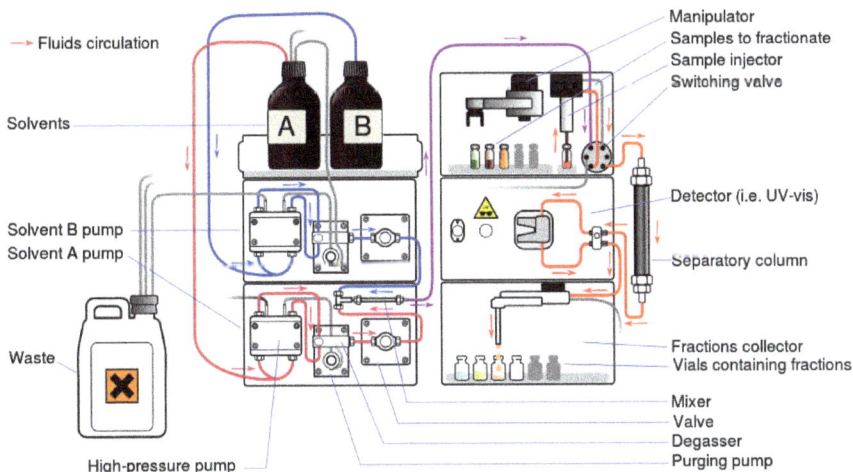

Preparative HPLC apparatus

In HPLC the sample is forced by a liquid at high pressure (the mobile phase) through a column that is packed with a stationary phase composed of irregularly or spherically shaped particles, a porous monolithic layer, or a porous membrane. HPLC is historically divided into two different sub-classes based on the polarity of the mobile and stationary phases. Methods in which the stationary phase is more polar than the mobile phase (e.g., toluene as the mobile phase, silica as the stationary phase) are termed normal phase liquid chromatography (NPLC) and the opposite (e.g., water-methanol mixture as the mobile phase and C18 (octadecylsilyl) as the stationary phase) is termed reversed phase liquid chromatography (RPLC).

Specific techniques under this broad heading are listed below.

Affinity Chromatography

Affinity chromatography is based on selective non-covalent interaction between an analyte and specific molecules. It is very specific, but not very robust. It is often used in biochemistry in the purification of proteins bound to tags. These fusion proteins are labeled with compounds such as His-tags, biotin or antigens, which bind to the stationary phase specifically. After purification, some of these tags are usually removed and the pure protein is obtained.

Affinity chromatography often utilizes a biomolecule's affinity for a metal (Zn, Cu, Fe, etc.). Columns are often manually prepared. Traditional affinity columns are used as a preparative step to flush out unwanted biomolecules.

However, HPLC techniques exist that do utilize affinity chromatography properties. Immobilized Metal Affinity Chromatography (IMAC) is useful to separate aforementioned molecules based on the relative affinity for the metal (I.e. Dionex IMAC). Often these columns can be loaded with different metals to create a column with a targeted affinity.

Supercritical Fluid Chromatography

Supercritical fluid chromatography is a separation technique in which the mobile phase is a fluid above and relatively close to its critical temperature and pressure.

Techniques by Separation Mechanism

Ion Exchange Chromatography

Ion exchange chromatography (usually referred to as ion chromatography) uses an ion exchange mechanism to separate analytes based on their respective charges. It is usually performed in columns but can also be useful in planar mode. Ion exchange chromatography uses a charged stationary phase to separate charged compounds including anions, cations, amino acids, peptides, and proteins. In conventional methods the stationary phase is an ion exchange resin that carries charged functional groups that interact with oppositely charged groups of the compound to retain. Ion exchange chromatography is commonly used to purify proteins using FPLC.

Size-exclusion Chromatography

Size-exclusion chromatography (SEC) is also known as gel permeation chromatography (GPC) or gel filtration chromatography and separates molecules according to their size (or more accurately according to their hydrodynamic diameter or hydrodynamic volume). Smaller molecules are able to enter the pores of the media and, therefore, molecules are trapped and removed from the flow of the mobile phase. The average residence time in the pores depends upon the effective size of the analyte molecules. However, molecules that are larger than the average pore size of the packing are excluded and thus suffer essentially no retention; such species are the first to be eluted. It is generally a low-resolution chromatography technique and thus it is often reserved for the final, "polishing" step of a purification. It is also useful for determining the tertiary structure and quaternary structure of purified proteins, especially since it can be carried out under native solution conditions.

Expanded Bed Adsorption Chromatographic Separation

An expanded bed chromatographic adsorption (EBA) column for a biochemical separation process comprises a pressure equalization liquid distributor having a self-cleaning function below a porous blocking sieve plate at the bottom of the expanded bed, an upper part nozzle assembly having a backflush cleaning function at the top of the expanded bed, a better distribution of the feedstock liquor added into the expanded bed ensuring that the fluid passed through the expanded bed layer displays a state of piston flow. The expanded bed layer displays a state of piston flow. The expanded bed chromatographic separation column has advantages of increasing the separation efficiency of the expanded bed.

Expanded-bed adsorption (EBA) chromatography is a convenient and effective technique for the capture of proteins directly from unclarified crude sample. In EBA chromatography, the settled bed is first expanded by upward flow of equilibration buffer. The crude feed, a mixture of soluble proteins, contaminants, cells, and cell debris, is then passed upward through the expanded bed. Target proteins are captured on the adsorbent, while particulates and contaminants pass through. A change to elution buffer while maintaining upward flow results in desorption of the target protein in expanded-bed mode. Alternatively, if the flow is reversed, the adsorbed particles will quickly settle and the proteins can be desorbed by an elution buffer. The mode used for elution (expanded-bed versus settled-bed) depends on the characteristics of the feed. After elution, the adsorbent is cleaned with a predefined cleaning-in-place (CIP) solution, with cleaning followed by either column regeneration (for further use) or storage.

Special Techniques

Reversed-phase Chromatography

Reversed-phase chromatography (RPC) is any liquid chromatography procedure in which the mobile phase is significantly more polar than the stationary phase. It is so named because in normal-phase liquid chromatography, the mobile phase is significantly less polar than the stationary phase. Hydrophobic molecules in the mobile phase tend to adsorb to the relatively hydrophobic stationary phase. Hydrophilic molecules in the mobile phase will tend to elute first. Separating columns typically comprise a C8 or C18 carbon-chain bonded to a silica particle substrate.

Hydrophobic Interaction Chromatography

Hydrophobic interactions between proteins and the chromatographic matrix can be exploited to purify proteins. In hydrophobic interaction chromatography the matrix material is lightly substituted with hydrophobic groups. These groups can range from methyl, ethyl, propyl, octyl, or phenyl groups. At high salt concentrations, non-polar sidechains on the surface on proteins "interact" with the hydrophobic groups; that is, both types of groups are excluded by the polar solvent (hydrophobic effects are augmented by increased ionic strength). Thus, the sample is applied to the column in a buffer which is highly polar. The eluant is typically an aqueous buffer with decreasing salt concentrations, increasing concentrations of detergent (which disrupts hydrophobic interactions), or changes in pH.

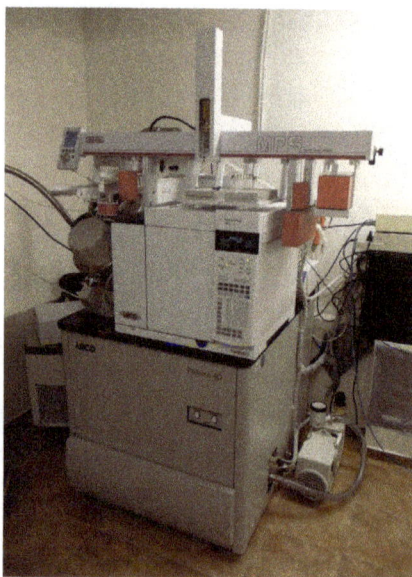

Two-dimensional chromatograph GCxGC-TOFMS at Chemical Faculty of GUT Gdańsk, Poland, 2016

In general, Hydrophobic Interaction Chromatography (HIC) is advantageous if the sample is sensitive to pH change or harsh solvents typically used in other types of chromatography but not high salt concentrations. Commonly, it is the amount of salt in the buffer which is varied. In 2012, Müller and Franzreb described the effects of temperature on HIC using Bovine Serum Albumin (BSA) with four different types of hydrophobic resin. The study altered temperature as to effect the binding affinity of BSA onto the matrix. It was concluded that cycling temperature from 50 degrees to 10 degrees would not be adequate to effectively wash all BSA from the matrix but could be very effective if the column would only be used a few times. Using temperature to effect change allows labs to cut costs on buying salt and saves money.

If high salt concentrations along with temperature fluctuations want to be avoided you can use a more hydrophobic to compete with your sample to elute it. [source] This so-called salt independent method of HIC showed a direct isolation of Human Immunoglobulin G (IgG) from serum with satisfactory yield and used Beta-cyclodextrin as a competitor to displace IgG from the matrix. This largely opens up the possibility of using HIC with samples which are salt sensitive as we know high salt concentrations precipitate proteins.

Two-dimensional Chromatography

In some cases, the chemistry within a given column can be insufficient to separate some analytes. It is possible to direct a series of unresolved peaks onto a second column with different physico-chemical (Chemical classification) properties. Since the mechanism of retention on this new solid support is different from the first dimensional separation, it can be possible to separate compounds that are indistinguishable by one-dimensional chromatography. The sample is spotted at one corner of a square plate,developed, air-dried, then rotated by 90° and usually redeveloped in a second solvent system.

Simulated Moving-bed Chromatography

The simulated moving bed (SMB) technique is a variant of high performance liquid chromatography; it is used to separate particles and/or chemical compounds that would be difficult or impossible to resolve otherwise. This increased separation is brought about by a valve-and-column arrangement that is used to lengthen the stationary phase indefinitely. In the moving bed technique of preparative chromatography the feed entry and the analyte recovery are simultaneous and continuous, but because of practical difficulties with a continuously moving bed, simulated moving bed technique was proposed. In the simulated moving bed technique instead of moving the bed, the sample inlet and the analyte exit positions are moved continuously, giving the impression of a moving bed. True moving bed chromatography (TMBC) is only a theoretical concept. Its simulation, SMBC is achieved by the use of a multiplicity of columns in series and a complex valve arrangement, which provides for sample and solvent feed, and also analyte and waste takeoff at appropriate locations of any column, whereby it allows switching at regular intervals the sample entry in one direction, the solvent entry in the opposite direction, whilst changing the analyte and waste takeoff positions appropriately as well.

Pyrolysis Gas Chromatography

Pyrolysis gas chromatography mass spectrometry is a method of chemical analysis in which the sample is heated to decomposition to produce smaller molecules that are separated by gas chromatography and detected using mass spectrometry.

Pyrolysis is the thermal decomposition of materials in an inert atmosphere or a vacuum. The sample is put into direct contact with a platinum wire, or placed in a quartz sample tube, and rapidly heated to 600–1000°C. Depending on the application even higher temperatures are used. Three different heating techniques are used in actual pyrolyzers: Isothermal furnace, inductive heating (Curie Point filament), and resistive heating using platinum filaments. Large molecules cleave at their weakest points and produce smaller, more volatile fragments. These fragments can be separated by gas chromatography. Pyrolysis GC chromatograms are typically complex because a wide range of different decomposition products is formed. The data can either be used as fingerprint to prove material identity or the GC/MS data is used to identify individual fragments to obtain structural information. To increase the volatility of polar fragments, various methylating reagents can be added to a sample before pyrolysis.

Besides the usage of dedicated pyrolyzers, pyrolysis GC of solid and liquid samples can be performed directly inside Programmable Temperature Vaporizer (PTV) injectors that provide quick heating (up to 30°C/s) and high maximum temperatures of 600–650°C. This is sufficient for some pyrolysis applications. The main advantage is that no dedicated instrument has to be purchased and pyrolysis can be performed as part of routine GC analysis. In this case quartz GC inlet liners have to be used. Quantitative data can be acquired, and good results of derivatization inside the PTV injector are published as well.

Fast Protein Liquid Chromatography

Fast protein liquid chromatography (FPLC), is a form of liquid chromatography that is often used to analyze or purify mixtures of proteins. As in other forms of chromatography, separation is possible because the different components of a mixture have different affinities for two materials, a moving fluid (the "mobile phase") and a porous solid (the stationary phase). In FPLC the mobile phase is an aqueous solution, or "buffer". The buffer flow rate is controlled by a positive-displacement pump and is normally kept constant, while the composition of the buffer can be varied by drawing fluids in different proportions from two or more external reservoirs. The stationary phase is a resin composed of beads, usually of cross-linked agarose, packed into a cylindrical glass or plastic column. FPLC resins are available in a wide range of bead sizes and surface ligands depending on the application.

Countercurrent Chromatography

Countercurrent chromatography (CCC) is a type of liquid-liquid chromatography, where both the stationary and mobile phases are liquids. The operating principle of CCC equipment requires a column consisting of an open tube coiled around a bobbin. The bobbin is rotated in a double-axis gyratory motion (a cardioid), which causes a variable gravity (G) field to act on the column during each rotation. This motion causes the column to see one partitioning step per revolution and components of the sample separate in the column due to their partitioning coefficient between the two immiscible liquid phases used. There are many types of CCC available today. These include HSCCC (High Speed CCC) and HPCCC (High Performance CCC). HPCCC is the latest and best performing version of the instrumentation available currently.

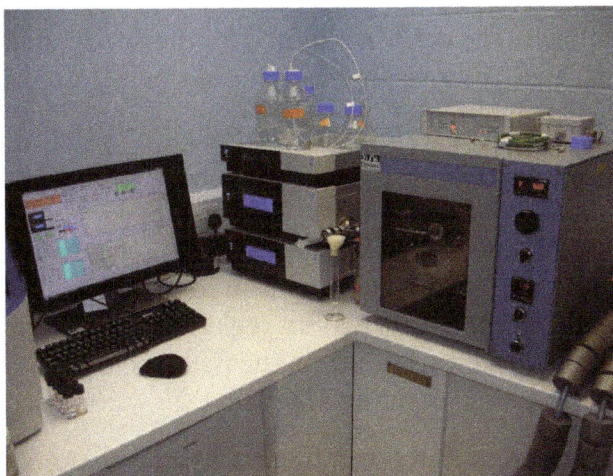

An example of a HPCCC system

Chiral Chromatography

Chiral chromatography involves the separation of stereoisomers. In the case of enantiomers, these have no chemical or physical differences apart from being three-dimensional mirror images. Conventional chromatography or other separation processes are incapable of separating them. To enable chiral separations to take place, either the mobile phase or the stationary phase must themselves be made chiral, giving differing affinities between the analytes. Chiral chromatography HPLC columns (with a chiral stationary phase) in both normal and reversed phase are commercially available.

Antibody

Lab Experiment: Preparation and Isolation of Polyclonal Antibody in Rabbit

Reagents and Equipments

1. Reagents for SDS-PAGE and agarose

2. Freund's incomplete adjuvant

3. Freund's complete adjuvant

4. Ethanol

5. Disposable Syringes

6. Eppendorf

7. Centrifuge

1.   Preparation of antigen- The antigen required for the development of polyclonal an-
      tibodies is ~2mg. it is required for multiple injections to induce robust immune
      response. It has following steps.

A.   Production of antigen: The most common method to produce antigen for antibody
      generation is utilization of recombinant DNA technology. An Over-view of steps
      involved in cloning protein of interest in given in thefigure below. In general, for
      biotechnology related application cloning is used to produce DNA, either as a part
      of a functional gene or part of regulatory region such as promoter. An outline of
      basic steps involved in cloning is given in the figure below. It has multiple steps to
      achieve cloned gene in a vector for amplification.

   •   Isolation or amplification of gene fragment from genome of the organism.

   •   Restriction digestion of gene fragment and vector

   •   Ligation of cut gene product and vector

   •   Insertion of ligated DNA or recombinant DNA into the host.

   •   Screening and selection of cells containing recombinant DNA.

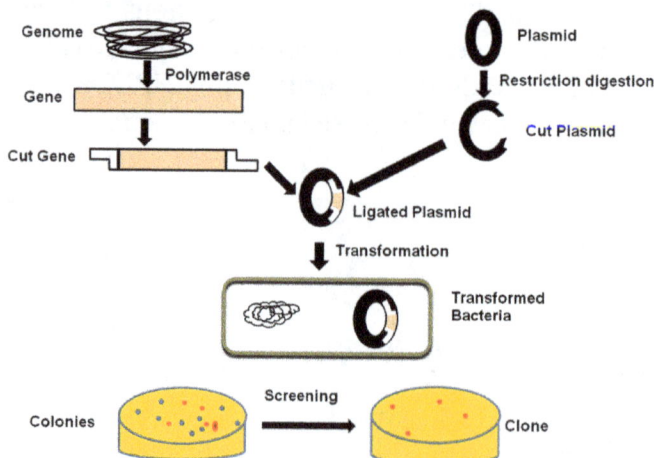

An over-view of different steps involved in cloning.

- Expression of gene using E. coli expression system- The steps in an expression of a gene is outlined in the figure and it has following steps:

  ◇ Transformation- As discussed we can use multiple methods to transform the host with a recombinant clone containing suitable selection marker.

  ◇ Growth of the bacteria- A single colony of the transformed colony is inoculated in the suitable media as discussed before upto the log phase (OD=0.6-0.7).

  ◇ Induction- The bacterial culture is now induced with IPTG (0-1mM) for 3-6 hrs to produce the protein.

  ◇ Recovery of the bacteria and analysis of protein expression- Bacteria can be recovered from the culture with a brief centrifugation at 8000-9000 RPM and analyzed on a SDS-PAGE. The SDS-PAGE analysis of a particular expression study in *E. coli* is given in the figure and it indicates a prominent expression of the target protein in the induced cells as compare to the uninduced cells.

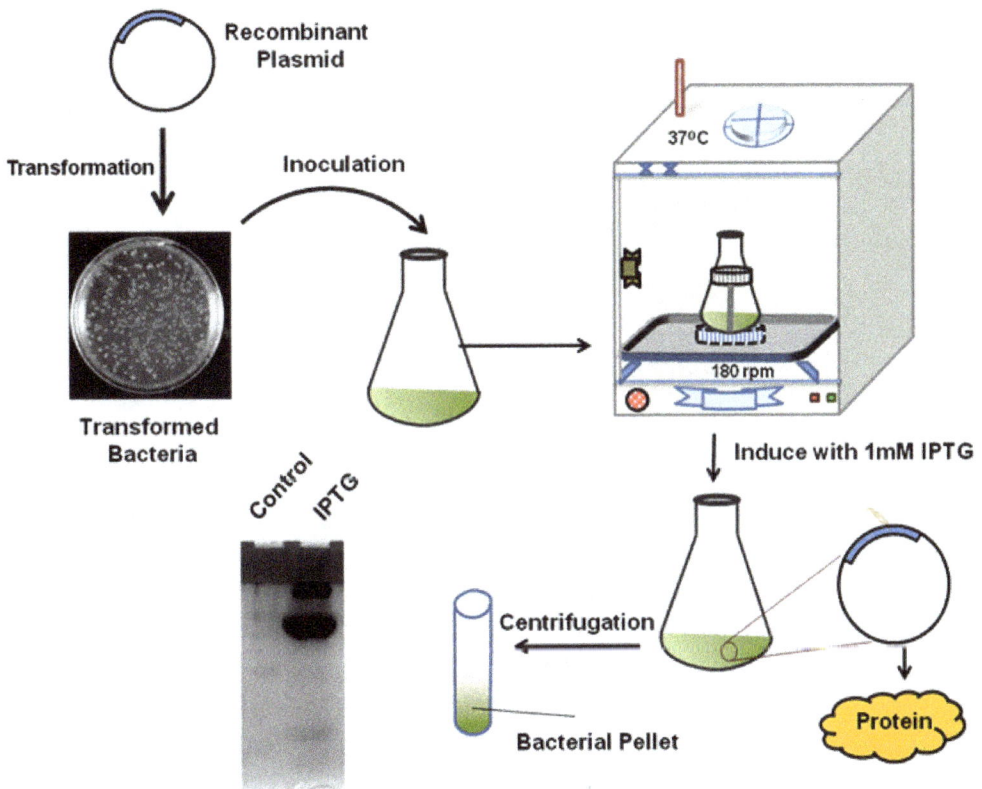

B. Isolation of antigen- There are two different approaches to isolate the antigen from the E. coli over-expressing cells.

B1. Purification of antigen under native conditions- The antigen can be expressed with a terminal affinity tag such as GST linker. The schematic to purify the protein wth affinity tag is given in the figure below. Glutathione S-transferase (GST) utilizes glutathione as a substrate to catalyze conjugation reactions for xenobiotic detoxification purposes. The recombinant fusion protein contains GST as a tag is purified with glutathione coupled matrix. GST fusion protein is produced by the recombining protein of interest with the GST coding sequence present in the expression vector (either before or after coding sequence of protein of interest). It is transformed, over-expressed and the bacterial lysate containing fusion protein is purified, using affinity column. The sample is loaded on the column previously equiliberated with the buffer containing high salt (0.5M NaCl). Unbound protein is washed with the equilibration buffer and then the fusion proein is eluted with different concentration of glutathione dissolved in the equilibration buffer. Purified fusion can be treated with the thrombin to remove the GST tag from the protein of interest. The mixture containing free GST tag and the protein can be purified using the affinity column again as tag will bind to the matrix but protein will come out in the unbound fraction.

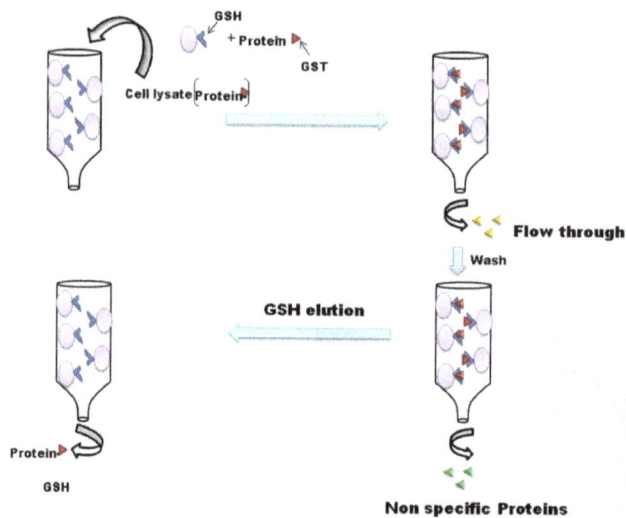

Purification of Antigen under native conditions

B2. Isolation of antigen under denaturating conditions- The schematic diagram to isolate the antigen under denaturating condition is given inthe figure below. In the electroelution, a gel band is cut from the SDS-PAGE and placed in a dialysis bag and sealed from both ends. The dialysis bag is choosen so that the molecular weight cut off of dialysis membrane should be lower than the protein of interest. The dialysis bag is placed in the horizontal gel apparatus with buffer and electrophoresis is performed with a constant voltage. During electrophoresis the protein band migrate and ultimately comes out from the gel block. Due to dialysis bag, salt and other small molecule contaminant moves out of the dialysis bag but protein remain trapped within the dialysis bag. Protein can be recovered from the dialysis bag for further use in downstream processing.

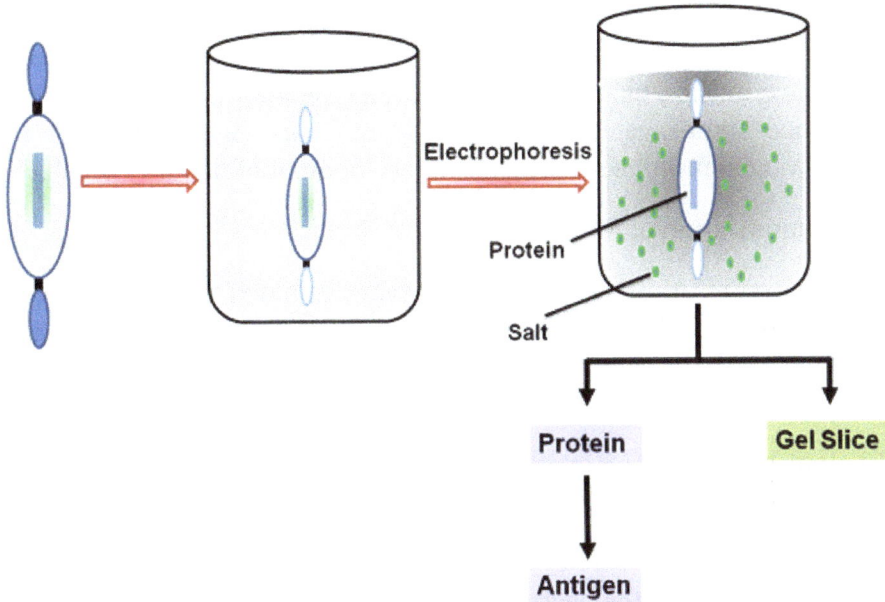

Electroelution using horizontal gel electrophoresis apparatus.

Procedure : The details procedure to isolate the antigen from the SDS-PAGE is given in thefigure below.

i)   Antigen band was excised from SDS polyacrylamide gel and placed in dialysis bag filled with 5-7 ml of SDS PAGE buffer.

ii) The dialysis bag was placed horizontally in horizontal electrophoresis tank.

iii) The protein was electro eluted from the gel using horizontal electrophoresis run at 80-100 volt overnight under cold condition to avoid heating.

iv) The gel piece was removed from the bag and the eluted protein was dialyzed to remove SDS and then lyophilized.

v)   The eluted protein was checked in SDS PAGE and Western blotting using tag specific antibody.

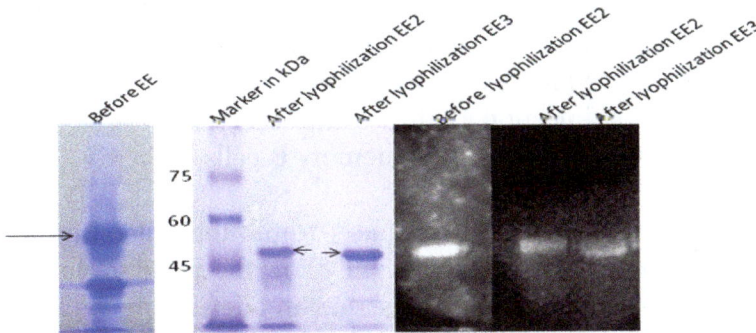

Analysis of Electroelution.

2. Preparation of Immunogen- Combine 100µl of antigen (100-150µg) with an equal volume of freund's incomplete adjuvant to a final volume of 200µl. Mix thoroughly to obtain the emulsion using a syringe or a pipette. after 4 weeks of first injection, inject first booster dosage. Repeat booster injection 4-5 times after every 4 weeks to generate a robust immune response and development of memory B-cells.

3. In-Vivo Immunization of Rabbit

   i.   Before immunization, take out 0.1-0.5ml mice blood from the tail vein before the first injection. Incubate the sample at 4°C at 30mins and allow the blood to clot. Centrifuge the sample at 7000g for 10min. Collect the serum and store it at -20°C and labeled as pre-immune serum.

   ii. Take out 5 mice (BALB/c strain) from the cage and sterile them by spraying 70% alchol. Inject 200µl antigen mixture per rabbit. During this step either use a helper to hold the rabbit or use a restrain device to hold the rabbit.

   iii. inject the antigen on the back of the rabbit in the form of buttons.

Rabbit used for immunization.

4. Booster : Combine 100µl of antigen (100-150µg) with an equal volume of freund's incomplete adjuvant to a final volume of 200µl. Mix thoroughly to obtain the emulsion using a syringe or a pipette. after 4 weeks of first injection, inject first booster dosage. Repeat booster injection 4-5 times after every 4 weeks to generate a robust immune response and development of memory B-cells.

5. Determination of Antibody Titre-Take out 5-10ml rabbit blood from the ear vein. Incubate the sample at 4 0 C at 30mins and allow the blood to clot. Centrifuge the sample at 7000g for 10min. Collect the serum and determine the antibody by a indirect ELISA (discussed in detail later).

6. Collection of blood and preparation of serum- Take out 20-30ml rabbit blood from the ear vein or large quantity of blood can be drawn after cardiac puncture (cardiac puncture is a terminal event and it is not recommended as rabbit will not survive for future immunization). Incubate the sample at 4 0 C at 30mins and allow the blood to clot. Centrifuge the sample at 7000g for 10min. Collect the serum and determine the antibody by a indirect ELISA (discussed in detail later).

Lab Experiment: Generation of Hybridoma and Isolate the Monoclonal Antibodies from the Clone

Background Information: Monoclonal antibody are produced from the single clone of B-Cells. The general procedure involved in generation and production of monoclonal antibodies is given in the figure below.

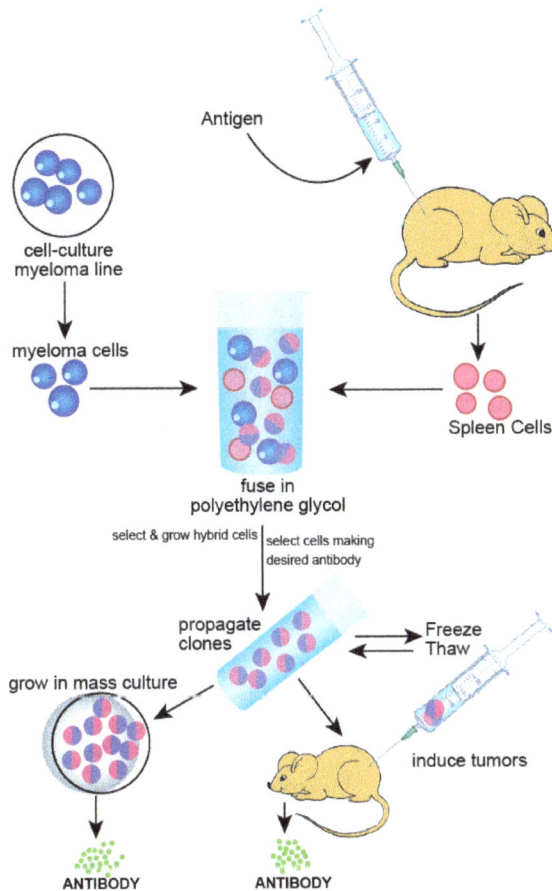

Different steps involved in monoclonal antibody generation and production

1. Purification of antigen- The antigen used to immunize be as pure as possible. Use of pure antigen reduces the generation of cross reactive antibodies.

2. Preparation of Immunogen- Combine 100µl of antigen (100-150µg) with an equal volume of freund's complete adjuvant to a final volume of 200µl. Mix thoroughly to

obtain the emulsion using a syringe or a pipette. The perfect emulsion of the antigen can be tested by dropping a small amount into the beaker containing water. A good emulsion will not spread on water surface.

3. In-Vivo Immunization of mice-

    i.   Before immunization, take out 0.1-0.5ml mice blood from the tail vein before the first injection. Incubate the sample at 4°C at 30mins and allow the blood to clot. Centrifuge the sample at 7000g for 10min. Collect the serum and store it at -20°C and labeled as pre-immune serum.

    ii. Take out 5 mice (BALB/c strain) from the cage and sterile them by spraying 70% alchol. Inject 200µl antigen mixture per mice. During this step either use a helper to hold the mice or use a restrain device to hold the mice. Briefly clean the injection site with 70% ethanol and inject antigen through multiple routes:

         a.   intravenous- Antigen mixture can be directly injected into the tail vein.

         b.   Intraperitoneal injections- While making i.p. injection avoid injecting the antigen into the stomatch.

         c.   Sub-cutaneous and intramuscula r injection into the tight muscle.

    iii. After injection, keep the mice back to their cage.

    iv. Combine 100µl of antigen (100-150µg) with an equal volume of freund's incomplete adjuvant to a final volume of 200µl. Mix thoroughly to obtain the emulsion using a syringe or a pipette. after 4 weeks of first injection, inject first booster dosage. Repeat booster injection 4-5 times after every 4 weeks to generate a robust immune response and development of memory B-cells.

4. Determination of Antibody Titre- take out 0.1-0.5ml mice blood from the tail vein before the first injection. Incubate the sample at 4°C at 30mins and allow the blood to clot. Centrifuge the sample at 7000g for 10min. Collect the serum and determine the antibody by a indirect ELISA (discussed in detail later).

5. Prepration of peritoneal excude cells- Peritonel excude cells (PECs) derived from the bovine hypothalamus as a feeder cells for culturing of hybridoma cells. It has following steps:

    i)   Sacrifice the non-immunized mice either by cervical dislocation or CO_2 asphyxiation.

    ii) Soak the dead mice in a beaker containing 95% ethanol prior to start the dissection of mice in the laminar hood.

    iii) Make small cut at the abdominal region and expose the peritoneal cavity. Inject

3-5ml of serum free DMEM into the peritoneal cavity using disposable syringe.

iv) Flush the peritoneum and collect the peritoneum excecude cells and plate it to the disposable petridish with 10ml serum free DMEM media.

v) Count the cells and dilute the cells to 4×10^5 cells/ml. Allow the cells to incubate for 2 days and possible contamination needs to checked before using these cells for hybridoma culture.

6. Preparation of Spleen cells-

i) Sacrifice the non-immunized mice either by cervical dislocation or CO_2 asphyxiation.

ii) Soak the dead mice in a beaker containing 95% ethanol prior to start the dissection of mice in the laminar hood.

iii) Make small cut at the abdominal region and dissect to remove spleen using forceps and place it in the disposable petridish.

iv) Inject 2-5ml serum free DMEM into the spleen and this step will swell the tissue.

v) Tease the tissue with the help of forcep and released the cells into the petridish. Remove the debris and cell clump.

vi) Centrifuge at 50g for 5-10 mins at RT. Incubate the cells in the grey's hemolytic solution [8ml of grey's A solution and 2ml grey's B solution]. This step will remove the RBCs from the spleen cells leaving myeloma cells.

vii) Collect the cells and plate in the T-flask and grow upto the mid log phase.

viii) Resuspend the cells in the serum free DMEM media at RT.

7. Fusion of spleen and myeloma cells-

i) Mix spleen and myeloma cells in a ratio of 5:1 oe 10:1 in a sterile centrifuge tube.

ii) Centrifuge the cells at 120g for 5min at RT and remove the supernatant.

iii) Gentle tap the bottom of the tube and add 1ml of 50% PEG 6000 in serum free media. PEG solution should be added drop wise to avoid clumping of the cells.

iv) Dilute the mixture by adding 3ml warm serum free DMEM over a period of 1-2 mins.

v) Centrifuge the fused cells and resuspend the cells in DMEM containing 20% FCS at a cell density 10^5 or 10^6 cells/ml.

vi) Add 50μl feeder PEC cells in 96 wells and on top of this add 50μl of fused cells. Incubate the cells at 37°C with 5% CO_2 for 24hrs prior to goto next step of screening these hybridomas.

8. Selection of hybridoma cells- Hybridoma cells contains spleen cells capable of coding antibody whereas myeloma cells doesn't produce antibodies but has ability to grow indefinitely. Incubating hybridoma cells in HAT medium allows growth of fused cells where as unfused individual myeloma cells or kidney cells doesnot be able to grow. The selection process has following steps:

i) Add 50ml HAT in complete DMEM media containing 20% FCS for 3days. Replace the media everyday with this solution.

ii) Check for colonies and possible contamination.

iii) Once the colonies are observed, isolate these cells by serial dilution method. The setup for isolating individual colony is given in the figure. Delineate the boundry of each colony with a marker from the back side of the plate. Remove the media and put cloning ring to each colony. Wash the colony with PBS and add 100μl trypsin-EDTA to remove the colony. wash the colony with PBS and transfer into one well of 24 well dish. Allow it to grow and become 80% confluent. Transfer these cells to the 6 well dish in the presence of selection media and allow it to reach 80% confluency. Take a small aliquot of the cell and test the presence of antibody with ELISA.

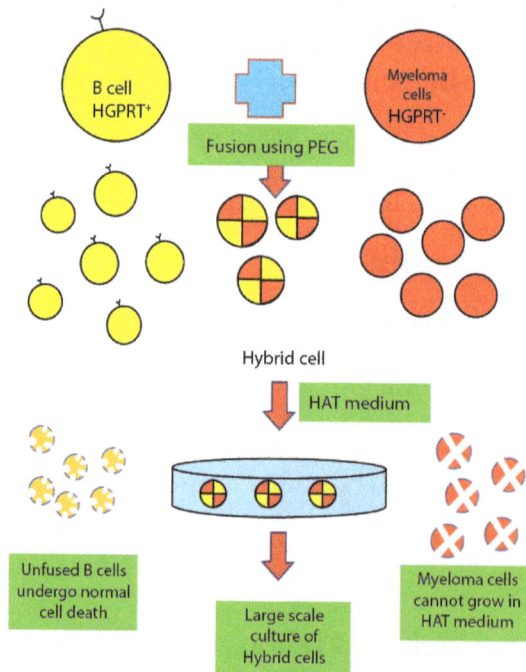

Different steps involved in hybridoma selection

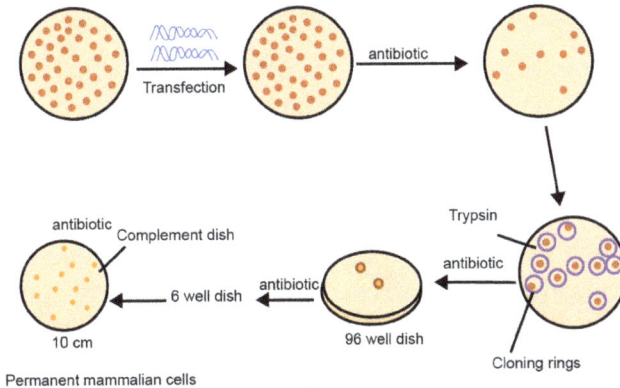

Different Steps in selection and isolation of hybridoma cells.

Molecular Mechanism of Hybridoma selection: Nucleic acid synthesis is essential for growth amd multiplication. HAT medium inhibits denovo pathways and cells with salvage pathways (containing HGPRT enzyme) can survive. Myeloma cells depends solely on denovo pathways for nuleic acid synthesis where as kidney cells has both denovo and salvage pathways. In the presence of HAT, individual myeloma cells or kidney cells will not survive where as only hybrid cells can be able to survive.

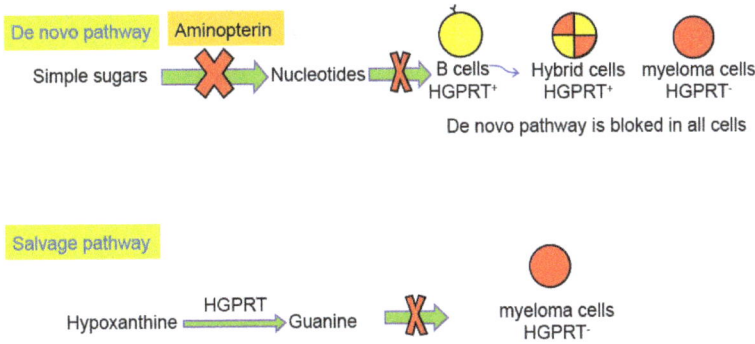

Molecular basis of hybridoma selection in HAT medium.

9. Screening of hybridoma supernatant for presence of antibody- The presence of antibody in the culture supernant is done by ELISA.

10. Harvesting of monoclonal antibody-

    i)   Once the color of the medium changes from red to yellow-orange, change the medium with the fresh DMEM containing 20% FCS, 1xHAT and 20% PECs.

    ii) The healthy cell lines that produces antibodies is transferred to the 24 well dishes containing 30-60ml of complete medium for large production of antibody.

    iii) Harvest the supernatant by transferring culture into a tube and centrifuge it at 120g for 10min at RT. Transfer the supernatant to fresh tube and adjust the pH 7.2. Add 0.1% sodium azide and preserve the supernatant at -20°C.

Lab Experiment: Generation of Affinity Column for Antibody Purification

Generation of Receptor- The receptor molecule present on the matrix can be produced either by genetic engineering, isolation from the crude extract or in the case of antibody, it is produced in the mouse/rabbit model and purify.

Coupling of the Receptor- Once the receptor molecule is available, it can be couple to the matrix by following steps. (1) Matrix activation (2) covalent coupling utilizing reactive group on ligand. (3) deactivation of the remaining active group on matrix.

CNBr mediated receptor coupling- CNBr mediated coupling is more suitable for protein/peptide to the polysaccharide matrix such as agarose or dextran. CNBr reacts with polysaccharide at pH 11-12 to form reactive cynate ester with matrix or less reactive cyclic imidocarbonate group. Under alkaline condition these cynogen ester reacts with the amine group on receptor to form isourea derivative. The amount of cynate ester is more with agarose whereas imidacarbonate is more formed with dextran as a matrix. The protein or peptide ligand with free amino group is added to the activated matrix to couple the receptor for affinity purification.

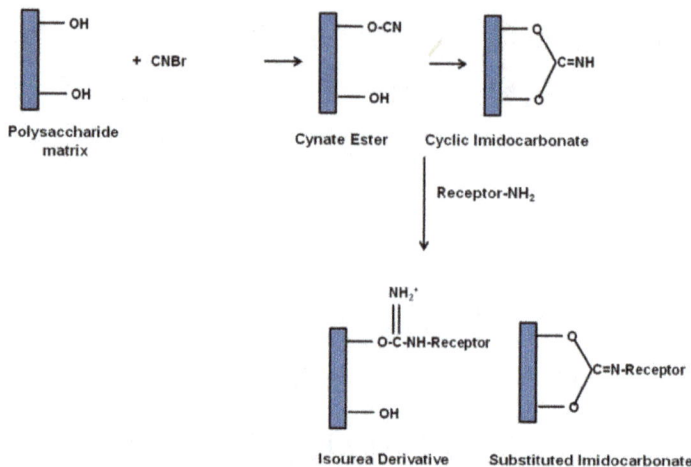

CNBr mediated coupling of receptor to the matrix.

Lab Experiment: Purification of Monoclonal Antibody Produced from the Hydridoma cell Lines using Affinity Chromatography

Operation of the Affinity chromatography- Different steps in affinity chromatography is given in the figure.

1. Equlibration- Affinity column material packed in a column and equilibrate with a buffer containing high salt (0.5M NaCl) to reduce the non-specific interaction of protein with the analyte.

2. Sample Preparation- The sample is prepared in the mobile phase and it should be free of suspended particle to avoid clogging of the column. The most recommended method to apply the sample is to inject the sample with a syringe. Load the supernatant from the hybridoma cell culture onto the column. Sample can reload onto the column 2-3 times to ensure 100% binding.

3. Wash the column 2 times with 10 colum volume using the equilibration buffer.

4. Elution- There are many ways to elute a analyte from the affinity column. (1) increasing concentration of counter ligand, (2) changing the pH polarity of the mobile phase, (3) By a detergent or chaotrophic salt to partially denature the receptor to reduce the affinity for bound ligand.

5. Neutrilize the acidic elute with 1M Tris pH 7.2 containing 150mM NaCl. The purified antigen can be stoed at -20°C.

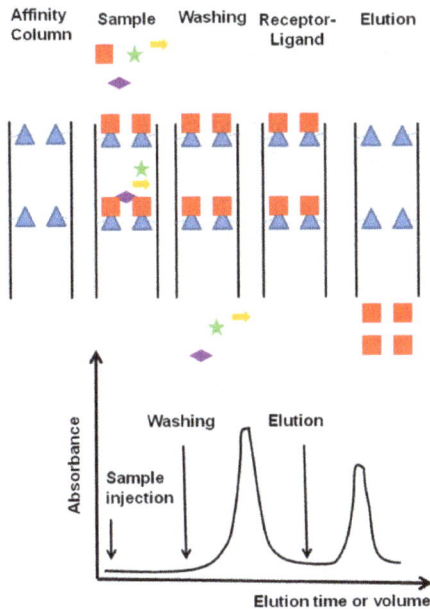

Performing Affinity chromatography

6. Column Regeneration- After the elution of analyte, affinity column requires a regeneration step to use next time. column is washed with 6M urea or guanidine hydrochloride to remove all non-specifically bound protein. The column is then equiliberated with mobile phase to regenerate the column. The column can be store at 4°C in the presence of 20% alchol containing 0.05% sodium azide.

References

- Minde DP (2012). "Determining biophysical protein stability in lysates by a fast proteolysis assay, FASTpp". PLOS ONE. 7 (10): e46147. doi:10.1371/journal.pone.0046147. PMC 3463568. PMID 23056252

- Edward I. Solomon; A. B. P. Lever (3 February 2006). Inorganic electronic structure and spectroscopy. Wiley-Interscience. p. 78. ISBN 978-0-471-97124-5. Retrieved 29 April 2011

- Song, Di; Ma, Shang; Khor, Soo Peang (2002-01-01). "Gel electrophoresis-autoradiographic image analysis of radiolabeled protein drug concentration in serum for pharmacokinetic studies". Journal of Pharmacological and Toxicological Methods. 47 (1): 59–66. ISSN 1056-8719. PMID 12387940

- Whitmore L, Wallace BA (2008). "Protein secondary structure analyses from circular dichroism spectroscopy: methods and reference databases". Biopolymers. 89 (5): 392–400. PMID 17896349. doi:10.1002/bip.20853

- Gerald D. Fasman (1996). Circular dichroism and the conformational analysis of biomolecules. Springer. pp. 3–. ISBN 978-0-306-45142-3. Retrieved 29 April 2011

- Hanaor, D.A.H.; Michelazzi, M.; Leonelli, C.; Sorrell, C.C. (2012). "The effects of carboxylic acids on the aqueous dispersion and electrophoretic deposition of ZrO_2". Journal of the European Ceramic Society. 32 (1): 235–244. doi:10.1016/j.jeurceramsoc.2011.08.015

- Greenfield NJ (2006). "Using circular dichroism spectra to estimate protein secondary structure". Nature Protocols. 1 (6): 2876–90. PMC 2728378. PMID 17406547. doi:10.1038/nprot.2006.202

- Kōji Nakanishi; Nina Berova; Robert Woody (1994). Circular dichroism: principles and applications. VCH. p. 473. ISBN 978-1-56081-618-8. Retrieved 29 April 2011

- Derrick, M.R., Stulik, D. and Landry J.M., Infrared Spectroscopy in Conservation Science, Scientific Tools for Conservation, Getty Publications, 2000. Retrieved December 11, 2015

- Soran Shadman; Charles Rose; Azer P. Yalin (2016). "Open-path cavity ring-down spectroscopy sensor for atmospheric ammonia". Applied Physics B. 122: 194. doi:10.1007/s00340-016-6461-5

- Alison Rodger; Bengt Nordén (1997). Circular dichroism and linear dichroism. Oxford University Press. ISBN 978-0-19-855897-2. Retrieved 29 April 2011

- N. Demirdöven; C. M. Cheatum; H. S. Chung; M. Khalil; J. Knoester; A. Tokmakoff (2004). "Two-dimensional infrared spectroscopy of antiparallel beta-sheet secondary structure". Journal of the American Chemical Society. 126 (25): 7981–90. doi:10.1021/ja049811j. PMID 15212548

- Booth, F. (1948). "Theory of Electrokinetic Effects". Nature. 161 (4081): 83–86. Bibcode:1948Natur.161...83B. PMID 18898334. doi:10.1038/161083a0

- Paula, Peter Atkins, Julio de (2009). Elements of physical chemistry (5th ed.). Oxford: Oxford U.P. p. 459. ISBN 978-0-19-922672-6

- Hoover, Rachel (February 21, 2014). "Need to Track Organic Nano-Particles Across the Universe? NASA's Got an App for That". NASA. Retrieved February 22, 2014

- Raymond S, Weintraub L (1959). "Acrylamide gel as a supporting medium for zone electrophoresis.". Science. 130 (3377): 711. PMID 14436634. doi:10.1126/science.130.3377.711

- Laurence M. Harwood; Christopher J. Moody (1989). Experimental organic chemistry: Principles and Practice (Illustrated ed.). Wiley-Blackwell. p. 292. ISBN 0-632-02017-2

- "What Happened When We Took the SCiO Food Analyzer Grocery Shopping". IEEE Spectrum: Technology, Engineering, and Science News. Retrieved 2017-03-23

5

Applied Areas of Biotechnology

Biotechnology has had an effect on human life and its advancement. A few of scientific progresses in the field of biotechnology are genetic engineering, insect control, herbicide resistant plants, resistance protein etcetera. It has further contributed in the field of medicine. Biotechnology's role in medicine which includes production of therapeutically important proteins, gene therapy and monoclonal antibody production among many others are also explored. The aspects elucidated in this chapter are of vital importance, and provide a better understanding of biotechnology.

Biotechnology in Plant Sciences

Biotechnology has influenced human life in many ways by invention to make his life more comfortable or easy. It has made advancement in agriculture to increase the yield, increases the value of animals by improving their breed and contributed significantly in the development and design of drugs against infectious diseases.

Biotechnology in plant sciences- Genetic Engineering has allowed us to produce genetically modified plants with diversified properties such as resistance against pest, drought, abiotic stress. These are few selected examples of advancement in the plant sciences due to technological contributions of biotechnology.

Insect control- Insect uses plants for nutrition and reduces the crop yield. There are two ways to control the effect of insects on the crop; reduction in the number of insects in the affected area or generation of plants with insect resistantance.

Sterile male insects- The mechanism of reducing the number of insects in the affected area through the use of sterile male insect is given in the figure below. In a typical insect life-cycle, male and female mate with each other to produce large number of fertilized eggs. Eggs go through a series of development stages to produce large number of baby insects to continue the life-cycle. In this approach, male insects are exposed to the radiation or other treatment in the laboratory to render them infertile. These sterile male insects are spread over the infected area. In the field, female mate with these sterile males but no offspring is produced. As a result over the course of time, the insect population will be reduced. The classical example of this approach is irradication of boll weevil, an insect responsible for the loss of cotton crop in USA.

Insect resistant plants- A genetically altered crop is produced to develop resistance against insects. One of the approach is to genetically modify the plant which will express a toxin to kill the insects but will be safe for human consumption. *Bacillus thuringiensis* (Bt) is a bacteria which secretes a insecticidal toxin. Spraying Bt toxin was in circulation to control the insect population. With the use of genetic engineering transgenic plants are produced which express Bt toxin in their somatic cells. When insect feeds on the plant, toxin reach to the stomatch and causes internal bleeding to kill the insect.

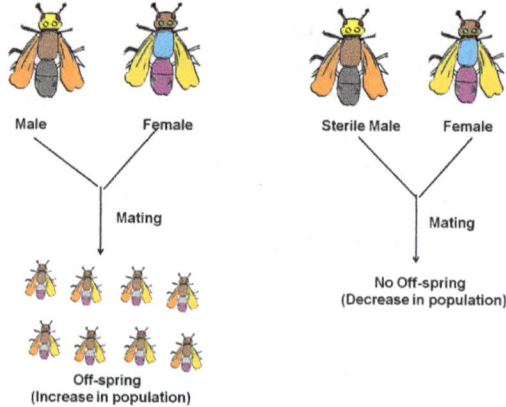

Schematic depiction of the impact of sterile male insects in controlling insect population.

Herbicide resistant plants- Weeds grow very fast and they compete for nutrients with the crop plant. Chemical herbicides are used in the agriculture to eradicate weeds from the fields. If weeds need to be removed from the crop, herbicide should do little or no effect on the crop plants. Herbicides are either selective towards a class of plant or non-selective to kill all plants they applied to and used more often to kill all vegetation. Glyphosate is one of the first herbicide designed to kill weeds. It interferes biosynthesis of aromatic amino acid tyrosine, phenylalanine and tryptophan by inhibiting enzyme 5-enolpyruvyl-shikimate-3-phosphate synthase (EPSP). The enzyme catalyzes the conversion of the shikimate-3-phosphate to the 5-enolpyruvylshikimate-3-phosphate. The treated plant can not be able to produce these amino acids as well as protein needed and dies. There are two approaches, adopted to develop herbicide resistantance in crop plant. (1) The genetically modified crop plant is designed with an alternate pathway to supply the aromatic amino acid to compensate the inhibition of EPSP. (2) Few bacterial strains use an alternate form of EPSP that is resistant to the glyphosate inhibition. The modified version of EPSP gene was isolated from the *Agrobacterium strain* CP4 and cloned into the crop plant to provide herbicide resistantance. So far the crop plant commercially available with herbicide resistantance are soy, maize, sorghum, canola and cotton.

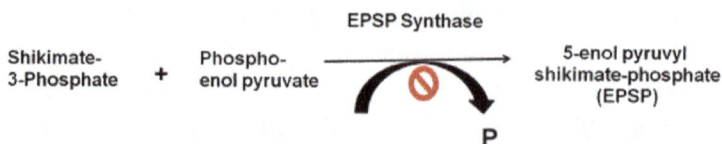

Biochemical reaction catalyzed by EPSP Synthase, a target of Glyphosate action.

Disease resistant plants- Plants are under continuous exposure to the pathogenic organism and the environmental conditions. Pathogenic organisms (bacteria, fungi, mycoplasma, virus) attack on plants to gain nutrients for their growth and disturb its metabolism to exhibit pathological symptoms. There are multiple approaches to develop disease resistant plant, although in few cases it is not possible to develop a disease resistant plant at all.

Selection and breeding of natural disease resistant plant species- Few naturally occurring plant species have acquired resistance against a particular disease. These species are preferred over other species for production. In few cases plant species resistant to the disease are either suspectible to other disease or the yield is low. In both cases, it is preferred that the disease resistant plant species can be cross breeding with a high yield plant species to acquire resistance as well as high yield.

Production of Resistance Protein- Plants have R gene (resistance gene) which produces R protein and these virulence factors allow acquiring resistance to combat pathogens. Every R gene recongnizes pathogen protein in a receptor-ligand fasion and as a result R gene product provides resistance against a particular pathogen or a family of related pathogens. R gene has the ability to modify its product to acquire resistance against new species of pathogen. A good example include barley MLO against powder mildew, wheat Lr34 against leaf rust, and wheat Yr36 against stripe rust.

Crop	Variety	Resistance
French Bean	1. Aigullion	Mosaic virus and anthracnose
	2. Hilda	Mosaic Virus
Broad bean	Futura	Chocolate spot
Cabbage	Stone head F1	Mildew
Carrot	Fly Away F1	Carrot fly
Cumcumber	Bush Chambion F1	Cucumber mosaic virus
Pea	Ambassador	Powdery mildew, fusarium wilt
Potato	1. Colleen	Blight and scab
	2. Osprey	Scab and eelworm
	3. Milva	Blight Resistant
Pepper	Bell Boy F1	Mosaic Virus resistant
Sweet Corn	Golden Sweet F1	Smut
Tomato	Alicante	Greenback and mildew

Selected List of disease resistant plant species.

Abiotic stress resistant plants- Over-production of systemin and HypSys has been found to provide resistance in plant against salt and UV radition. In the transgenic plants, systemin lower down the stomatal opening in comparison to the normal plants to reduce the loss of water. Whereas in higher salt solution, plants had larger stomotal opening, lower concentration of abscisic acid and proline.

Biotechnology in Animal Breeding

Biotechnology in animal breeding- Biotechnology has greatly facilitates the animal breeding and improving their species with additional traits.

Artificial insemination- The over-all process involves the introduction of male sperm into the reproductive tract of the female animal artificially. The availability of superior breed animal is due to the artificial insemination (AI). There are several advantages of AI compare to the natural breeding.

1. The male of a high breed (commonly known as sires) is very costly in comparison to the semen from them.

2. Through a natural breeding process, many diseases can be transmitted to the female through mating. These possibilities are much reduced in an AI procedure.

3. The high breed animal imported from other countries needs to go through quarantine process to ensure no spreading of disease. This process is costly and time comsuming.

Steps in artificial insemination- In an AI procedure, semen is collected from the superior breed male in a test tube. The quality of the sperm such as motility, number is checked through a microscopic observation. It is mixed with the extenders such as milk, yolk, glycerine and antibiotics and stored in liquid nitrogen for future use. Before go for final step of AI, it is important to study the estrus cycle of the female to know the exact time of ovulation. The semen is injected into the female reproductive tract to facilitate the process of fertilization. A skilled technician is required to perform most of the steps of artificial insemination.

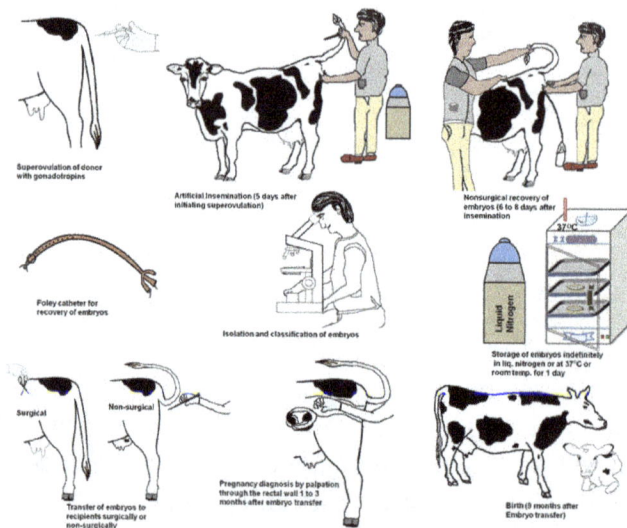

Steps of embryo transfer between donor and recipient.

Embryo Transfer- A superior breed female (commonly known as dam) can be able to give more offspring in a year by the process of embryo transfer. In this process, using an AI procedure superior breed male semen is used to fertilize the eggs in the female. Once the embryo is formed, it is transferred in another female (low breed) to produce offspring. The process of embryo transfer is given in the figure. The donor female is treated with the hormone to produce several eggs, the process known as superovulation. The egg is fertilized either by natural intercourse or by artificial insemination. The fertilized eggs are recovered from the uterus of the donor animal using a catheter. The solution from catheter is used to flush the uterus and fallopian tubes. On average 6-8 embryos are collected in each flush. The quality of the embryo is checked in the microscope and suitable embryo is implanted into the recipient animal. It is important that both donor and recipient animals must have a synchronized estrus cycle for successful embryo transfer.

Biotechnology in Medicine

Biotechnology in medicine- medicines are class of molecules used to correct the disturbances in the host physiology. They can be chemical in nature and used to inhibit aberrant enzymatic activity from the host or pathogen. In few cases, host enzymes can be supplied as a drug formulation to drive the biochemical reaction. Biotechnology has potentials in contributing into the development of the drug molecules. Besides this, biotechnology contributes many ways into the medicine field.

(A) Production of Therapeutically Important Proteins

A large number of genetic or metabolic diseases can be corrected by the supplying proteins or factors. Following the advancement in the biotechnology, many other proteins or factor produced are produced in different bacterial expression systems. In an approach, gene of the enzyme or proteinous factor is cloned into the appropriate plasmid to produce recombinant clone. For example, production of human insulin. Insulin is a dimer of an A chain and B-chain linked by disulphide bonds, composed of 51 amino acids with a molecular weight of 5808. A schematic presentation of steps in insulin production is given in the figure below. In this process, gene A and B is cloned into the bacterial plasmid separately to produce two recombinant clones. Peptide chain A and B is over-expressed in the E. coli and recombined together to produce functional insulin.

Other important examples of production of proteinous factors are as follows-

(I) Recombinant Chymosin- chymosin is required for manufacturing cheese. A non-pathogenic E. coli strain (K-12) is used for large scale production. The recombinant enzyme is safe to use, cost less and available in abundance.

(II) Recombinant Human Growth hormone (HGH)- It is produced by the pituitary gland and hormone is required to support growth and development of human. Recombinant HGH is cheap and safe to use for therapeutic applications.

(III) Recombinant blood clotting factor VIII- In a normal individual, blood loss from a damaged blood vessel is prevented by the formation of a clot. Blood clotting is a series of reaction involving different factors. Factor VIII is deficient in bleeding disorder such as hemophilia, and recombinant factor VIII is supplied to improve the disease condition.

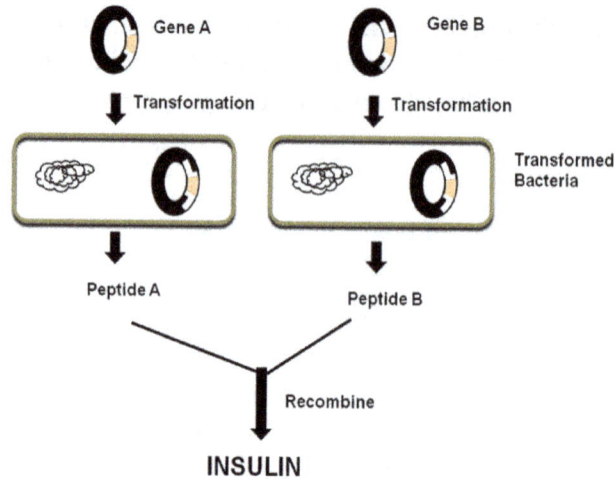

Recombinant DNA Technology to produce human insulin.

(B) Gene Therapy

As discussed before, production and supply of recombinant proteins is a temporarily solution for the treatment of a disease condition. In another approach, human expression system is used to produce the proteinous factor after inserting the recombinant clone into the human cells or inside the human body. Recombinant DNA is packed into the appropriate DNA delivery system (either a virus or liposome mediated) to deliver the gene into the human cells to correct the mutated genes or encode a therapeutic protein drug to provide treatment.

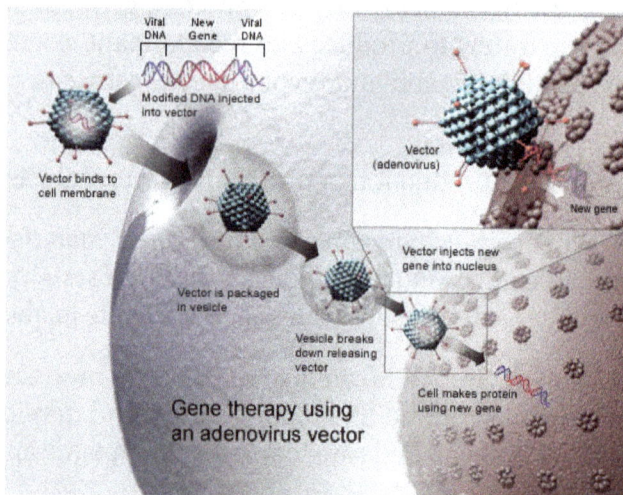

Gene therapy using an adenovirus vector. In some cases, the adenovirus will insert the new gene into a cell. If the treatment is successful, the new gene will make a functional protein to treat a disease.

Gene therapy is the therapeutic delivery of nucleic acid polymers into a patient's cells as a drug to treat disease. The first attempt at modifying human DNA was performed in 1980 by Martin Cline, but the first successful nuclear gene transfer in humans, approved by the National Institutes of Health, was performed in May 1989. The first therapeutic use of gene transfer as well as the first direct insertion of human DNA into the nuclear genome was performed by French Anderson in a trial starting in September 1990.

Between 1989 and February 2016, over 2,300 clinical trials had been conducted, more than half of them in phase I.

It should be noted that not all medical procedures that introduce alterations to a patient's genetic makeup can be considered gene therapy. Bone marrow transplantation and organ transplants in general have been found to introduce foreign DNA into patients. Gene therapy is defined by the precision of the procedure and the intention of direct therapeutic effects.

Background

Gene therapy was conceptualized in 1972, by authors who urged caution before commencing human gene therapy studies.

The first attempt, an unsuccessful one, at gene therapy (as well as the first case of medical transfer of foreign genes into humans not counting organ transplantation) was performed by Martin Cline on 10 July 1980. Cline claimed that one of the genes in his patients was active six months later, though he never published this data or had it verified and even if he is correct, it's unlikely it produced any significant beneficial effects treating beta-thalassemia.

After extensive research on animals throughout the 1980s and a 1989 bacterial gene tagging trial on humans, the first gene therapy widely accepted as a success was demonstrated in a trial that started on 14 September 1990, when Ashi DeSilva was treated for ADA-SCID.

The first somatic treatment that produced a permanent genetic change was performed in 1993.

This procedure was referred to sensationally and somewhat inaccurately in the media as a "three parent baby", though mtDNA is not the primary human genome and has little effect on an organism's individual characteristics beyond powering their cells.

Gene therapy is a way to fix a genetic problem at its source. The polymers are either translated into proteins, interfere with target gene expression, or possibly correct genetic mutations.

The most common form uses DNA that encodes a functional, therapeutic gene to replace a mutated gene. The polymer molecule is packaged within a "vector", which carries the molecule inside cells.

Early clinical failures led to dismissals of gene therapy. Clinical successes since 2006 regained researchers' attention, although as of 2014, it was still largely an experimental technique. These include treatment of retinal diseases Leber's congenital amaurosis and choroideremia, X-linked SCID, ADA-SCID, adrenoleukodystrophy, chronic lymphocytic leukemia (CLL), acute lymphocytic leukemia (ALL), multiple myeloma, haemophilia and Parkinson's disease. Between 2013 and April 2014, US companies invested over $600 million in the field.

The first commercial gene therapy, Gendicine, was approved in China in 2003 for the treatment of certain cancers. In 2011 Neovasculgen was registered in Russia as the first-in-class gene-therapy drug for treatment of peripheral artery disease, including critical limb ischemia. In 2012 Glybera, a treatment for a rare inherited disorder, became the first treatment to be approved for clinical use in either Europe or the United States after its endorsement by the European Commission.

Approaches

Following early advances in genetic engineering of bacteria, cells, and small animals, scientists started considering how to apply it to medicine. Two main approaches were considered – replacing or disrupting defective genes. Scientists focused on diseases caused by single-gene defects, such as cystic fibrosis, haemophilia, muscular dystrophy, thalassemia and sickle cell anemia. Glybera treats one such disease, caused by a defect in lipoprotein lipase.

DNA must be administered, reach the damaged cells, enter the cell and either express or disrupt a protein. Multiple delivery techniques have been explored. The initial approach incorporated DNA into an engineered virus to deliver the DNA into a chromosome. Naked DNA approaches have also been explored, especially in the context of vaccine development.

Generally, efforts focused on administering a gene that causes a needed protein to be expressed. More recently, increased understanding of nuclease function has led to more direct DNA editing, using techniques such as zinc finger nucleases and CRISPR. The vector incorporates genes into chromosomes. The expressed nucleases then knock out and replace genes in the chromosome. As of 2014 these approaches involve removing cells from patients, editing a chromosome and returning the transformed cells to patients.

Gene editing is a potential approach to alter the human genome to treat genetic diseases, viral diseases, and cancer. As of 2016 these approaches were still years from being medicine.

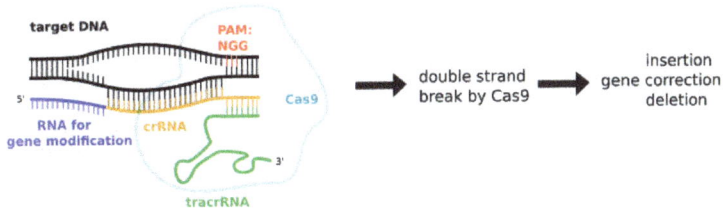

A duplex of crRNA and tracrRNA acts as guide RNA to introduce a specifically located gene modification based on the RNA 5' upstream of the crRNA. Cas9 binds the tracrRNA and needs a DNA binding sequence (5'NGG3'), which is called protospacer adjacent motif (PAM). After binding, Cas9 introduces a DNA double strand break, which is then followed by gene modification via homologous recombination (HDR) or non-homologous end joining (NHEJ).

Cell Types

Gene therapy may be classified into two types:

Somatic

In somatic cell gene therapy (SCGT), the therapeutic genes are transferred into any cell other than a gamete, germ cell, gametocyte or undifferentiated stem cell. Any such modifications affect the individual patient only, and are not inherited by offspring. Somatic gene therapy represents mainstream basic and clinical research, in which therapeutic DNA (either integrated in the genome or as an external episome or plasmid) is used to treat disease.

Over 600 clinical trials utilizing SCGT are underway in the US. Most focus on severe genetic disorders, including immunodeficiencies, haemophilia, thalassaemia and cystic fibrosis. Such single gene disorders are good candidates for somatic cell therapy. The complete correction of a genetic disorder or the replacement of multiple genes is not yet possible. Only a few of the trials are in the advanced stages.

Germline

In germline gene therapy (GGT), germ cells (sperm or eggs) are modified by the introduction of functional genes into their genomes. Modifying a germ cell causes all the organism's cells to contain the modified gene. The change is therefore heritable and passed on to later generations. Australia, Canada, Germany, Israel, Switzerland and the Netherlands prohibit GGT for application in human beings, for technical and ethical reasons, including insufficient knowledge about possible risks to future generations and higher risks versus SCGT. The US has no federal controls specifically addressing human genetic modification (beyond FDA regulations for therapies in general).

Vectors

The delivery of DNA into cells can be accomplished by multiple methods. The two major classes are recombinant viruses (sometimes called biological nanoparticles or viral vectors) and naked DNA or DNA complexes (non-viral methods).

Viruses

In order to replicate, viruses introduce their genetic material into the host cell, tricking the host's cellular machinery into using it as blueprints for viral proteins. Scientists exploit this by substituting a virus's genetic material with therapeutic DNA. (The term 'DNA' may be an oversimplification, as some viruses contain RNA, and gene therapy could take this form as well.) A number of viruses have been used for human gene therapy, including retrovirus, adenovirus, lentivirus, herpes simplex, vaccinia and adeno-associated virus. Like the genetic material (DNA or RNA) in viruses, therapeutic DNA can be designed to simply serve as a temporary blueprint that is degraded naturally or (at least theoretically) to enter the host's genome, becoming a permanent part of the host's DNA in infected cells.

Non-viral

Non-viral methods present certain advantages over viral methods, such as large scale production and low host immunogenicity. However, non-viral methods initially produced lower levels of transfection and gene expression, and thus lower therapeutic efficacy. Later technology remedied this deficiency.

Methods for non-viral gene therapy include the injection of naked DNA, electroporation, the gene gun, sonoporation, magnetofection, the use of oligonucleotides, lipoplexes, dendrimers, and inorganic nanoparticles.

Hurdles

Some of the unsolved problems include:

- Short-lived nature – Before gene therapy can become a permanent cure for a condition, the therapeutic DNA introduced into target cells must remain functional and the cells containing the therapeutic DNA must be stable. Problems with integrating therapeutic DNA into the genome and the rapidly dividing nature of many cells prevent it from achieving long-term benefits. Patients require multiple treatments.

- Immune response – Any time a foreign object is introduced into human tissues, the immune system is stimulated to attack the invader. Stimulating the immune system in a way that reduces gene therapy effectiveness is possible. The immune system's enhanced response to viruses that it has seen before reduces the effectiveness to repeated treatments.

- Problems with viral vectors – Viral vectors carry the risks of toxicity, inflammatory responses, and gene control and targeting issues.

- Multigene disorders – Some commonly occurring disorders, such as heart disease, high blood pressure, Alzheimer's disease, arthritis, and diabetes, are affected by variations in multiple genes, which complicate gene therapy.

- Some therapies may breach the Weismann barrier (between soma and germ-line) protecting the testes, potentially modifying the germline, falling afoul of regulations in countries that prohibit the latter practice.

- Insertional mutagenesis – If the DNA is integrated in a sensitive spot in the genome, for example in a tumor suppressor gene, the therapy could induce a tumor. This has occurred in clinical trials for X-linked severe combined immunodeficiency (X-SCID) patients, in which hematopoietic stem cells were transduced with a corrective transgene using a retrovirus, and this led to the development of T cell leukemia in 3 of 20 patients. One possible solution is to add a functional tumor suppressor gene to the DNA to be integrated. This may be problematic since the longer the DNA is, the harder it is to integrate into cell genomes. CRISPR technology allows researchers to make much more precise genome changes at exact locations.

- Cost – Alipogene tiparvovec or Glybera, for example, at a cost of $1.6 million per patient, was reported in 2013 to be the world's most expensive drug.

Deaths

Three patients' deaths have been reported in gene therapy trials, putting the field under close scrutiny. The first was that of Jesse Gelsinger in 1999. One X-SCID patient died of leukemia in 2003. In 2007, a rheumatoid arthritis patient died from an infection; the subsequent investigation concluded that the death was not related to gene therapy.

History

1970s and Earlier

In 1972 Friedmann and Roblin authored a paper in *Science* titled "Gene therapy for human genetic disease?" Rogers (1970) was cited for proposing that *exogenous good DNA* be used to replace the defective DNA in those who suffer from genetic defects.

1980s

In 1984 a retrovirus vector system was designed that could efficiently insert foreign genes into mammalian chromosomes.

1990s

The first approved gene therapy clinical research in the US took place on 14 September 1990, at the National Institutes of Health (NIH), under the direction of William French

Anderson. Four-year-old Ashanti DeSilva received treatment for a genetic defect that left her with ADA-SCID, a severe immune system deficiency. The effects were temporary, but successful.

Cancer gene therapy was introduced in 1992/93 (Trojan et al. 1993). The treatment of glioblastoma multiforme, the malignant brain tumor whose outcome is always fatal, was done using a vector expressing antisense IGF-I RNA (clinical trial approved by NIH n° 1602, and FDA in 1994). This therapy also represents the beginning of cancer immunogene therapy, a treatment which proves to be effective due to the anti-tumor mechanism of IGF-I antisense, which is related to strong immune and apoptotic phenomena.

In 1992 Claudio Bordignon, working at the Vita-Salute San Raffaele University, performed the first gene therapy procedure using hematopoietic stem cells as vectors to deliver genes intended to correct hereditary diseases. In 2002 this work led to the publication of the first successful gene therapy treatment for adenosine deaminase-deficiency (SCID). The success of a multi-center trial for treating children with SCID (severe combined immune deficiency or "bubble boy" disease) from 2000 and 2002, was questioned when two of the ten children treated at the trial's Paris center developed a leukemia-like condition. Clinical trials were halted temporarily in 2002, but resumed after regulatory review of the protocol in the US, the United Kingdom, France, Italy and Germany.

In 1993 Andrew Gobea was born with SCID following prenatal genetic screening. Blood was removed from his mother's placenta and umbilical cord immediately after birth, to acquire stem cells. The allele that codes for adenosine deaminase (ADA) was obtained and inserted into a retrovirus. Retroviruses and stem cells were mixed, after which the viruses inserted the gene into the stem cell chromosomes. Stem cells containing the working ADA gene were injected into Andrew's blood. Injections of the ADA enzyme were also given weekly. For four years T cells (white blood cells), produced by stem cells, made ADA enzymes using the ADA gene. After four years more treatment was needed.

Jesse Gelsinger's death in 1999 impeded gene therapy research in the US. As a result, the FDA suspended several clinical trials pending the reevaluation of ethical and procedural practices.

2000s

The modified cancer gene therapy strategy of antisense IGF-I RNA (NIH n° 1602) using antisense / triple helix anti IGF-I approach was registered in 2002 by Wiley gene therapy clinical trial - n° 635 and 636. The approach has shown promising results in the treatment of six different malignant tumors: glioblastoma, cancers of liver, colon, prostate, uterus and ovary (Collaborative NATO Science Programme on Gene Therapy USA, France, Poland n° LST 980517 conducted by J. Trojan) (Trojan et al., 2012). This anti–gene antisense/triple helix therapy has proven to be efficient, due to the mechanism stopping simultaneously IGF-I expression on translation and transcription levels, strengthening anti-tumor immune and apoptotic phenomena.

2002

Sickle-cell disease can be treated in mice. The mice – which have essentially the same defect that causes human cases – used a viral vector to induce production of fetal hemoglobin (HbF), which normally ceases to be produced shortly after birth. In humans, the use of hydroxyurea to stimulate the production of HbF temporarily alleviates sickle cell symptoms. The researchers demonstrated this treatment to be a more permanent means to increase therapeutic HbF production.

A new gene therapy approach repaired errors in messenger RNA derived from defective genes. This technique has the potential to treat thalassaemia, cystic fibrosis and some cancers.

Researchers created liposomes 25 nanometers across that can carry therapeutic DNA through pores in the nuclear membrane.

2003

In 2003 a research team inserted genes into the brain for the first time. They used liposomes coated in a polymer called polyethylene glycol, which, unlike viral vectors, are small enough to cross the blood–brain barrier.

Short pieces of double-stranded RNA (short, interfering RNAs or siRNAs) are used by cells to degrade RNA of a particular sequence. If a siRNA is designed to match the RNA copied from a faulty gene, then the abnormal protein product of that gene will not be produced.

Gendicine is a cancer gene therapy that delivers the tumor suppressor gene p53 using an engineered adenovirus. In 2003, it was approved in China for the treatment of head and neck squamous cell carcinoma.

2006

In March researchers announced the successful use of gene therapy to treat two adult patients for X-linked chronic granulomatous disease, a disease which affects myeloid cells and damages the immune system. The study is the first to show that gene therapy can treat the myeloid system.

In May a team reported a way to prevent the immune system from rejecting a newly delivered gene. Similar to organ transplantation, gene therapy has been plagued by this problem. The immune system normally recognizes the new gene as foreign and rejects the cells carrying it. The research utilized a newly uncovered network of genes regulated by molecules known as microRNAs. This natural function selectively obscured their therapeutic gene in immune system cells and protected it from discovery. Mice infected with the gene containing an immune-cell microRNA target sequence did not reject the gene.

In August scientists successfully treated metastatic melanoma in two patients using killer T cells genetically retargeted to attack the cancer cells.

In November researchers reported on the use of VRX496, a gene-based immunotherapy for the treatment of HIV that uses a lentiviral vector to deliver an antisense gene against the HIV envelope. In a phase I clinical trial, five subjects with chronic HIV infection who had failed to respond to at least two antiretroviral regimens were treated. A single intravenous infusion of autologous CD4 T cells genetically modified with VRX496 was well tolerated. All patients had stable or decreased viral load; four of the five patients had stable or increased CD4 T cell counts. All five patients had stable or increased immune response to HIV antigens and other pathogens. This was the first evaluation of a lentiviral vector administered in a US human clinical trial.

2007

In May researchers announced the first gene therapy trial for inherited retinal disease. The first operation was carried out on a 23-year-old British male, Robert Johnson, in early 2007.

2008

Leber's congenital amaurosis is an inherited blinding disease caused by mutations in the RPE65 gene. The results of a small clinical trial in children were published in April. Delivery of recombinant adeno-associated virus (AAV) carrying RPE65 yielded positive results. In May two more groups reported positive results in independent clinical trials using gene therapy to treat the condition. In all three clinical trials, patients recovered functional vision without apparent side-effects.

2009

In September researchers were able to give trichromatic vision to squirrel monkeys. In November 2009, researchers halted a fatal genetic disorder called adrenoleukodystrophy in two children using a lentivirus vector to deliver a functioning version of ABCD1, the gene that is mutated in the disorder.

2010

An April paper reported that gene therapy addressed achromatopsia (color blindness) in dogs by targeting cone photoreceptors. Cone function and day vision were restored for at least 33 months in two young specimens. The therapy was less efficient for older dogs.

In September it was announced that an 18-year-old male patient in France with beta-thalassemia major had been successfully treated. Beta-thalassemia major is an inherited blood disease in which beta haemoglobin is missing and patients are dependent on regular lifelong blood transfusions. The technique used a lentiviral vector to transduce the human ß-globin gene into purified blood and marrow cells obtained from the patient in June 2007. The patient's haemoglobin levels were stable at 9 to 10 g/dL. About a third of the hemoglobin contained the form introduced by the viral vector and blood transfusions were not needed. Further clinical trials were planned. Bone marrow transplants are the only cure for thalassemia, but 75% of patients do not find a matching donor.

Cancer immunogene therapy using modified anti – gene, antisense / triple helix approach was introduced in South America in 2010/11 in La Sabana University, Bogota (Ethical Committee 14 December 2010, no P-004-10). Considering the ethical aspect of gene diagnostic and gene therapy targeting IGF-I, the IGF-I expressing tumors i.e. lung and epidermis cancers, were treated (Trojan et al. 2016).

2011

In 2007 and 2008, a man was cured of HIV by repeated hematopoietic stem cell transplantation with double-delta-32 mutation which disables the CCR5 receptor. This cure was accepted by the medical community in 2011. It required complete ablation of existing bone marrow, which is very debilitating.

In August two of three subjects of a pilot study were confirmed to have been cured from chronic lymphocytic leukemia (CLL). The therapy used genetically modified T cells to attack cells that expressed the CD19 protein to fight the disease. In 2013, the researchers announced that 26 of 59 patients had achieved complete remission and the original patient had remained tumor-free.

Human HGF plasmid DNA therapy of cardiomyocytes is being examined as a potential treatment for coronary artery disease as well as treatment for the damage that occurs to the heart after myocardial infarction.

In 2011 Neovasculgen was registered in Russia as the first-in-class gene-therapy drug for treatment of peripheral artery disease, including critical limb ischemia; it delivers the gene encoding for VEGF. Neovasculogen is a plasmid encoding the CMV promoter and the 165 amino acid form of VEGF.

2012

The FDA approved Phase 1 clinical trials on thalassemia major patients in the US for 10 participants in July. The study was expected to continue until 2015.

In July 2012, the European Medicines Agency recommended approval of a gene therapy treatment for the first time in either Europe or the United States. The treatment used Alipogene tiparvovec (Glybera) to compensate for lipoprotein lipase deficiency, which can cause severe pancreatitis. The recommendation was endorsed by the European Commission in November 2012 and commercial rollout began in late 2014.

In December 2012, it was reported that 10 of 13 patients with multiple myeloma were in remission "or very close to it" three months after being injected with a treatment involving genetically engineered T cells to target proteins NY-ESO-1 and LAGE-1, which exist only on cancerous myeloma cells.

2013

In March researchers reported that three of five adult subjects who had acute lymphocytic leukemia (ALL) had been in remission for five months to two years after being treated with genetically modified T cells which attacked cells with CD19 genes on their surface, i.e. all B-cells, cancerous or not. The researchers believed that the patients' immune systems would make normal T-cells and B-cells after a couple of months. They were also given bone marrow. One patient relapsed and died and one died of a blood clot unrelated to the disease.

Following encouraging Phase 1 trials, in April, researchers announced they were starting Phase 2 clinical trials (called CUPID2 and SERCA-LVAD) on 250 patients at several hospitals to combat heart disease. The therapy was designed to increase the levels of SERCA2, a protein in heart muscles, improving muscle function. The FDA granted this a Breakthrough Therapy Designation to accelerate the trial and approval process. In 2016 it was reported that no improvement was found from the CUPID 2 trial.

In July researchers reported promising results for six children with two severe hereditary diseases had been treated with a partially deactivated lentivirus to replace a faulty gene and after 7–32 months. Three of the children had metachromatic leukodystrophy, which causes children to lose cognitive and motor skills. The other children had Wiskott-Aldrich syndrome, which leaves them to open to infection, autoimmune diseases and cancer. Follow up trials with gene therapy on another six children with Wiskott-Aldrich syndrome were also reported as promising.

In October researchers reported that two children born with adenosine deaminase severe combined immunodeficiency disease (ADA-SCID) had been treated with genetically engineered stem cells 18 months previously and that their immune systems were showing signs of full recovery. Another three children were making progress. In 2014 a further 18 children with ADA-SCID were cured by gene therapy. ADA-SCID children have no functioning immune system and are sometimes known as "bubble children."

Also in October researchers reported that they had treated six haemophilia sufferers in early 2011 using an adeno-associated virus. Over two years later all six were producing clotting factor.

2014

In January researchers reported that six choroideremia patients had been treated with adeno-associated virus with a copy of REP1. Over a six-month to two-year period all had improved their sight. By 2016, 32 patients had been treated with positive results and researchers were hopeful the treatment would be long-lasting. Choroideremia is an inherited genetic eye disease with no approved treatment, leading to loss of sight.

In March researchers reported that 12 HIV patients had been treated since 2009 in a trial with a genetically engineered virus with a rare mutation (CCR5 deficiency) known to protect against HIV with promising results.

Clinical trials of gene therapy for sickle cell disease were started in 2014. There is a need for high quality randomised controlled trials assessing the risks and benefits invoved with gene therapy for people with sickle cell disease.

2015

In February LentiGlobin BB305, a gene therapy treatment undergoing clinical trials for treatment of beta thalassemia gained FDA "breakthrough" status after several patients were able to forgo the frequent blood transfusions usually required to treat the disease.

In March researchers delivered a recombinant gene encoding a broadly neutralizing antibody into monkeys infected with simian HIV; the monkeys' cells produced the antibody, which cleared them of HIV. The technique is named immunoprophylaxis by gene transfer (IGT). Animal tests for antibodies to ebola, malaria, influenza and hepatitis were underway.

In March scientists, including an inventor of CRISPR, urged a worldwide moratorium on germline gene therapy, writing "scientists should avoid even attempting, in lax jurisdictions, germline genome modification for clinical application in humans" until the full implications "are discussed among scientific and governmental organizations".

Also in 2015 Glybera was approved for the German market.

In October, researchers announced that they had treated a baby girl, Layla Richards, with an experimental treatment using donor T-cells genetically engineered using TALEN to attack cancer cells. One year after the treatment she was still free of her cancer (a highly aggressive form of acute lymphoblastic leukaemia [ALL]). Children with highly aggressive ALL normally have a very poor prognosis and Layla's disease had been regarded as terminal before the treatment.

In December, scientists of major world academies called for a moratorium on inheritable human genome edits, including those related to CRISPR-Cas9 technologies but that basic research including embryo gene editing should continue.

2016

In April the Committee for Medicinal Products for Human Use of the European Medicines Agency endorsed a gene therapy treatment called Strimvelis and the European Commission approved it in June. This treats children born with ADA-SCID and who have no functioning immune system - sometimes called the "bubble baby" disease. This was the second gene therapy treatment to be approved in Europe.

In October, Chinese scientists reported they had started a trial to genetically modify T-cells from 10 adult patients with lung cancer and reinject the modified T-cells back into their bodies to attack the cancer cells. The T-cells had the PD-1 protein (which stops or slows the immune response) removed using CRISPR-Cas9.

A 2016 Cochrane systematic review looking at data from four trials on topical cystic fibrosis transmembrane conductance regulator (CFTR) gene therapy does not support its clinical use as a mist inhaled into the lungs to treat cystic fibrosis patients with lung infections. One of the four trials did find weak evidence that liposome-based CFTR gene transfer therapy may lead to a small respiratory improvement for people with CF. This weak evidence is not enough to make a clinical recommendation for routine CFTR gene therapy.

2017

In February Kite Pharma announced results a clinical trial of CAR-T cells in around a hundred people with advanced Non-Hodgkin lymphoma.

In March, French scientists reported on clinical research of gene therapy to treat sickle-cell disease.

Speculative uses

Speculated uses for gene therapy include:

Fertility

Gene Therapy techniques have the potential to provide alternative treatments for those with infertility. Recently, successful experimentation on mice has proven that fertility can be restored by using the gene therapy method, CRISPR. Spermatogenical stem cells from another organism were transplanted into the testes of an infertile male mouse. The stem cells re-established spermatogenesis and fertility.

Gene Doping

Athletes might adopt gene therapy technologies to improve their performance. Gene doping is not known to occur, but multiple gene therapies may have such effects. Kayser et al. argue that gene doping could level the playing field if all athletes receive equal access. Critics claim that any therapeutic intervention for non-therapeutic/enhancement purposes compromises the ethical foundations of medicine and sports.

Human Genetic Engineering

Genetic engineering could be used to change physical appearance, metabolism, and even improve physical capabilities and mental faculties such as memory and intelligence. Ethical claims about germline engineering include beliefs that every fetus has a right to remain genetically unmodified, that parents hold the right to genetically modify their offspring, and that every child has the right to be born free of preventable diseases. For adults, genetic engineering could be seen as another enhancement technique to add to diet, exercise, education, cosmetics and plastic surgery. Another theorist claims that moral concerns limit but do not prohibit germline engineering.

Possible regulatory schemes include a complete ban, provision to everyone, or professional self-regulation. The American Medical Association's Council on Ethical and Judicial Affairs stated that "genetic interventions to enhance traits should be considered permissible only in severely restricted situations: (1) clear and meaningful benefits to the fetus or child; (2) no trade-off with other characteristics or traits; and (3) equal access to the genetic technology, irrespective of income or other socioeconomic characteristics."

As early in the history of biotechnology as 1990, there have been scientists opposed to attempts to modify the human germline using these new tools, and such concerns have continued as technology progressed. With the advent of new techniques like CRISPR, in March 2015 a group of scientists urged a worldwide moratorium on clinical use of gene editing technologies to edit the human genome in a way that can be inherited. In April 2015, researchers sparked controversy when they reported results of basic research to edit the DNA of non-viable human embryos using CRISPR. A committee of the American National Academy of Sciences and National Academy of Medicine gave qualified support to human genome editing in 2017 once answers have been found to safety and efficiency problems "but only for serious conditions under stringent oversight."

Regulations

Regulations covering genetic modification are part of general guidelines about human-involved biomedical research.

The Helsinki Declaration (Ethical Principles for Medical Research Involving Human Subjects) was amended by the World Medical Association's General Assembly in 2008. This document provides principles physicians and researchers must consider when involving humans as research subjects. The Statement on Gene Therapy Research initiated by the Human Genome Organization (HUGO) in 2001 provides a legal baseline for all countries. HUGO's document emphasizes human freedom and adherence to human rights, and offers recommendations for somatic gene therapy, including the importance of recognizing public concerns about such research.

United States

No federal legislation lays out protocols or restrictions about human genetic engineering. This subject is governed by overlapping regulations from local and federal agencies, including the Department of Health and Human Services, the FDA and NIH's Recombinant DNA Advisory Committee. Researchers seeking federal funds for an investigational new drug application, (commonly the case for somatic human genetic engineering), must obey international and federal guidelines for the protection of human subjects.

NIH serves as the main gene therapy regulator for federally funded research. Privately funded research is advised to follow these regulations. NIH provides funding for research that develops or enhances genetic engineering techniques and to evaluate the ethics and quality in current research. The NIH maintains a mandatory registry of human genetic engineering research protocols that includes all federally funded projects.

An NIH advisory committee published a set of guidelines on gene manipulation. The guidelines discuss lab safety as well as human test subjects and various experimental types that involve genetic changes. Several sections specifically pertain to human genetic engineering, including Section III-C-1. This section describes required review processes and other aspects when seeking approval to begin clinical research involving genetic transfer into a human patient. The protocol for a gene therapy clinical trial must be approved by the NIH's Recombinant DNA Advisory Committee prior to any clinical trial beginning; this is different from any other kind of clinical trial.

As with other kinds of drugs, the FDA regulates the quality and safety of gene therapy products and supervises how these products are used clinically. Therapeutic alteration of the human genome falls under the same regulatory requirements as any other medical treatment. Research involving human subjects, such as clinical trials, must be reviewed and approved by the FDA and an Institutional Review Board.

Popular culture

Gene therapy is the basis for the plotline of the film *I Am Legend* and the TV show *Will Gene Therapy Change the Human Race?*. It is also used in Stargate as a means of allowing humans to use Ancient technology.

References

- Ghosh, Pallab (28 April 2016). "Gene therapy reverses sight loss and is long-lasting". BBC News, Science & Environment. Retrieved 29 April 2016

- Malech, H. L.; Ochs, H. D. (2015). "An Emerging Era of Clinical Benefit from Gene Therapy". JAMA (Journal of the American Medical Association). 313 (15): 1522. doi:10.1001/jama.2015.2055

- Strachnan, T.; Read, A. P. (2004). Human Molecular Genetics (3rd ed.). Garland Publishing. p. 616. ISBN 0815341849

- Coghlan, Andy (11 December 2013) Souped-up immune cells force leukaemia into remission. New Scientist. Retrieved 15 April 2013

- Gallagher, James. (2 November 2012) BBC News – Gene therapy: Glybera approved by European Commission. BBC. Retrieved 15 December 2012

- Ledford, Heidi (12 October 2016). "CRISPR deployed to combat sickle-cell anaemia". Nature. Retrieved 13 October 2016

- Roco MC; Bainbridge WS (2002). "Converging Technologies for Improving Human Performance: Integrating From the Nanoscale". Journal of Nanoparticle Research. 4 (4): 281–295. doi:10.1023/A:1021152023349

- Grens, Kerry (13 October 2016). "CRISPR Corrects Sickle Cell-Causing Gene In Human Cells". The Scientist. Retrieved 10 October. 2016

- Allhoff, F. (2005). "Germ-Line Genetic Enhancement and Rawlsian Primary Goods". Kennedy Institute of Ethics Journal. 15 (1): 39–56. doi:10.1353/ken.2005.0007. PMID 15881795

- "The first gene therapy". Life Sciences Foundation. 21 June 2011. Archived from the original on 28 November 2012. Retrieved 7 January 2014

- Stein, Rob (11 October 2010). "First patient treated in stem cell study". The Washington Post. Retrieved 10 November 2010

PERMISSIONS

We would like to thank the editorial team for lending their expertise to make the book truly unique. They have played a crucial role in the development of this book. Without their invaluable contributions this book wouldn't have been possible. They have made vital efforts to compile up to date information on the varied aspects of this subject to make this book a valuable addition to the collection of many professionals and students.

This book was conceptualized with the vision of imparting up-to-date and integrated information in this field. To ensure the same, a matchless editorial board was set up. Every individual on the board went through rigorous rounds of assessment to prove their worth. After which they invested a large part of their time researching and compiling the most relevant data for our readers.

The editorial board has been involved in producing this book since its inception. They have spent rigorous hours researching and exploring the diverse topics which have resulted in the successful publishing of this book. They have passed on their knowledge of decades through this book. To expedite this challenging task, the publisher supported the team at every step. A small team of assistant editors was also appointed to further simplify the editing procedure and attain best results for the readers.

Apart from the editorial board, the designing team has also invested a significant amount of their time in understanding the subject and creating the most relevant covers. They scrutinized every image to scout for the most suitable representation of the subject and create an appropriate cover for the book.

The publishing team has been an ardent support to the editorial, designing and production team. Their endless efforts to recruit the best for this project, has resulted in the accomplishment of this book. They are a veteran in the field of academics and their pool of knowledge is as vast as their experience in printing. Their expertise and guidance has proved useful at every step. Their uncompromising quality standards have made this book an exceptional effort. Their encouragement from time to time has been an inspiration for everyone.

The publisher and the editorial board hope that this book will prove to be a valuable piece of knowledge for students, practitioners and scholars across the globe.

Index

A

Acrylamide Gels, 163
Agrobacterium Strain, 196
Allosteric Database, 30
Allosteric Regulation, 27-30
Allosteric Sites, 28, 30, 33
Alpha-ketoglutarate, 39, 41-42
Anaerobic Oxidation, 24, 46
Antibody, 60-61, 73, 75, 101, 181-182, 186-188, 190-192, 195, 211
Antibody Purification, 192
Artificial Chromosome, 53-54, 107, 109
Artificial Insemination, 5, 198-199

B

Bacterial Expression, 3, 47, 62-63, 199
Bacteriophage, 54, 89, 106
Bio-process Technology, 2
Biological Molecules, 140-141
Blood Transfusions, 209, 211
Bradford Assay, 117-120

C

Cancers, 100, 202, 206-207, 209
Carbohydrate Metabolism, 24, 44
Catabolism, 24, 35, 41
Cdna Library, 52, 55-57, 59-60, 105
Cellular Structure, 2
Chloroplast, 20, 100
Circular Dichroism, 129, 135-141, 143, 194
Circular Polarization, 135
Citric Acid Cycle, 33-35, 38, 40-43, 51
Cloning, 48, 52-54, 56-57, 69-71, 75-88, 92-94, 97, 103-111, 182, 190
Cosmid, 102-103, 106
Cytosol, 15, 17-19, 22, 25, 34, 42-43

D

Drug Targets, 33

E

Electrophoresis, 89, 94, 158-163, 165-166, 168, 184-185, 194
Embryo Gene, 212
Embryo Transfer, 82, 198-199
Endangered Species, 80, 83, 85-86
Episomal Vector, 108
Ethidium Bromide, 99, 165
Eukaryotic, 15, 17-18, 21, 24, 34, 55-56, 69, 108, 161
Eukaryotic Cells, 15, 24, 34, 55, 69, 108, 161

F

Fermenting Organism, 2
Fertility, 212

G

Gene Doping, 213
Gene Sequence, 52
Gene Therapy, 11, 195, 200-215
Genetic Engineering, 3-5, 11-13, 93, 192, 195-196, 202, 213-214
Genetic Make-up, 2
Germline, 205, 211, 213
Glycolysis, 19, 24-27, 31, 35, 38-41, 43-44, 46
Golgi Bodies, 22-23
Growth Media, 2, 47-49

H

Heavy Water, 14
Hybridoma, 187-191, 193
Hydrogen Bonding, 125

I

Immunological Methods, 60, 112
Infrared Spectroscopy, 143, 150-151, 153, 194
Integrating Vector, 109
Ir Spectroscopy, 143, 146, 150-151

L
Lifespan, 86
Lysosomes, 22-23

M
Mammalian Vector, 110
Metabolic Pathways, 2, 40, 44
Microbial Strains, 2
Mitochondria, 19-20, 22, 24, 34, 38, 80-81
Molar Ellipticity, 138-140
Monoclonal Antibody, 187, 191-192, 195
Morpheein Model, 29

N
Nucleus, 15-19, 55, 68-69, 79-81, 83, 86

O
Organic Chemistry, 194
Outer Flagella, 16
Ovalbumin, 119
Oxaloacetate, 35, 38, 40-42

P
Parthenogenesis, 81
Peptidoglycan Layer, 16-17
Pharmacology, 10
Photosynthesis, 21
Plasmid, 11, 17, 56, 61, 66-67, 69, 94, 103-110, 159, 199, 203, 209
Polarity, 176, 193
Polyacrylamide Gel Electrophoresis, 89, 159, 161, 165

P
Polymerase Chain Reaction, 70, 72, 75, 86, 95-97
Prokaryotic, 15-17, 24, 34, 56, 67, 88, 92-93, 108, 161
Protein Standard, 118-119

R
Recognition Site, 88-92
Recombinant Dna, 52, 57, 69-73, 75-76, 89, 110-111, 182, 200, 214
Reporter Gene, 105
Residue Ellipticity, 140
Resistance Gene, 17, 65, 104, 109, 197
Resistance Protein, 195, 197
Restriction Enzyme, 53-54, 71, 88-90, 93-94, 103

S
Sickle-cell Disease, 207
Spectroscopy, 112, 129-130, 133, 135, 140-144, 146, 148-153, 194
Stem Cells, 79, 205-206, 210, 212

T
Transformation Efficiency, 64, 66, 109
Transmembrane Conductance Regulator, 212

U
Unfolding Curve, 132-134
Uv Radiation, 113-115

V
Vibrational Modes, 145, 153
Viruses, 71, 74, 88, 95, 99-100, 111, 161, 203-204